T0247805

THE INNER CLOCK

THE INNER CLOCK

Living in Sync with Our Circadian Rhythms

LYNNE PEEPLES

RIVERHEAD BOOKS
NEW YORK
2024

RIVERHEAD BOOKS
An imprint of Penguin Random House LLC
penguinrandomhouse.com

Copyright © 2024 by Lynne Peeples
Penguin Random House supports copyright. Copyright fuels creativity, encourages diverse voices, promotes free speech, and creates a vibrant culture. Thank you for buying an authorized edition of this book and for complying with copyright laws by not reproducing, scanning, or distributing any part of it in any form without permission. You are supporting writers and allowing Penguin Random House to continue to publish books for every reader.

Riverhead and the R colophon are registered trademarks of Penguin Random House LLC.

LIBRARY OF CONGRESS CATALOGING-IN-PUBLICATION DATA
Names: Peeples, Lynne, author.
Title: The inner clock: living in sync with our circadian rhythms / Lynne Peeples.
Description: New York: Riverhead Books, 2024. | Includes bibliographical references. |
Identifiers: LCCN 2024013333 (print) | LCCN 2024013334 (ebook) | ISBN 9780593538906 (hardcover) | ISBN 9780593538920 (epub)
Subjects: LCSH: Circadian rhythms—Health aspects—Popular works. | Sleep-wake cycle—Popular works.
Classification: LCC QP84.6 .P44 2024 (print) | LCC QP84.6 (ebook) | DDC 612/.022—dc23/eng/20240510
LC record available at https://lccn.loc.gov/2024013333
LC ebook record available at https://lccn.loc.gov/2024013334

Printed in the United States of America
1st Printing

Book design by Daniel Lagin

PUBLISHER'S NOTE

Neither the publisher nor the author is engaged in rendering professional advice or services to the individual reader. The ideas, procedures, and suggestions contained in this book are not intended as a substitute for consulting with your physician. All matters regarding your health require medical supervision. Neither the author nor the publisher shall be liable or responsible for any loss or damage allegedly arising from any information or suggestion in this book.

For Dad, and in loving memory of Mom.

Thank you for giving me rhythm and showing me light.

Contents

Part III
RESET

Introduction

Just before the Coast Guard icebreaker *Polar Star* set off from my hometown of Seattle in December 2020, I sent its commanding officer an early Christmas present: a table lamp. Sure, that may sound on par with socks from Santa. But this was no ordinary lamp. The lamp Amazon overnighted to Captain William Woityra was specially designed to emulate the sun's bright bluish white light, one of the key cues the human body uses to tell time.

The COVID-19 pandemic had changed the course for Captain Woityra and his crew. Rather than their typical winter cruise south to support scientists in Antarctica—where sunlight never ceases and penguins are plentiful—they were now headed north to hang with polar bears in the unrelenting darkness of the Arctic. I had met Captain Woityra through a friend. At the time, as I still am today, I was telling anyone who would listen about the amazing little clocks inside our cells and the essential role daylight plays in keeping them ticking right—metaphorically, of course. I was both intrigued and concerned by the scant supply of sun that awaited the icebreaker crew. The gift was my crude attempt to help Captain Woityra keep his biological rhythms beating as steadily as possible over the next few months. I suggested he sit in front of the lamp for twenty to thirty minutes within the first couple of hours of waking up, a tip I'd gleaned from my early research.

I'm not sure I had Captain Woityra convinced, but he gamely placed the lamp on his desk aboard the 399-foot ship, beside a couple of bottles of Purell and beneath a wall-mounted navigational compass. As *Polar Star* departed, he at least feigned enthusiasm about the light's potential to boost his smarts, happiness, and health—and maybe even limit his visits to the onboard "Polar Starbucks."

At this very moment, a symphony of miniature timepieces is ticking throughout your body—in your stomach and skin, in your liver and lungs, even in the bones and muscles of your legs. It is directing you to feel everything you feel—hungry, sleepy, alert, strong—at just the right time. At a minimum, it wants you to be active during the day and restful in the evening. And it is constantly on the lookout for clues to ensure it remains in sync with the sun. Orchestrating this round-the-clock production, termed your circadian rhythm, is a master timekeeper in your brain.

Unfortunately, you don't have to travel to the North or South Pole to derail this life-sustaining system. Threats come from all directions. Artificial light, jet lag, contrived time zones, air pollution, late-night meals, and myriad other modern insults wreak havoc on our internal clocks, leaving them beating to different drums. This can sabotage sleep, reduce productivity, and raise risks of ailments: obesity, heart disease, hair loss, digestive disorders, and depression make the mounting list. Need another worry to keep you up at night? The consequences of circadian disruption may even be passed down to future generations.

I found it difficult at first to believe that mistimed clocks could cause so much damage. If circadian rhythms had such far-reaching impacts, why wasn't everyone talking about them? Why had I learned nothing about them in school? Maybe their tireless ticking had been too obvious for us to take notice?

Our physiology and behavior evolved to be highly rhythmic. And that regularity remains deeply, almost invisibly, embedded in our everyday lives and in the lives of the creatures around us. Nearly all life on Earth developed internal cycles to pair with the predictable patterns of the sun, its illumination of the sky, and other environmental cues, such as temperature and tides. Different species, from cyanobacteria to corn to cheetahs, leverage variations of this complex biological mechanism. Take a starling or bunting, for example. These birds can pick a flight course and hold to it by using their inner clocks to compensate for the reliable movement of the sun or constellations. Then there's us, with our clock-scrambling streetlights, smartphone screens, and shift-work schedules.

Too little light during the day and too much light at night leads the assault on our biological rhythms. As we live more of our hours indoors, we experience less contrast between day and night. This makes it harder for our bodies to distinguish the two. Countless other changes set our clocks apart from our ancestors and confound our clocks, from the rapidity with which we cross continents, to the pervasive contaminants tainting our environment, to the odd times we eat and exercise. Early humans traveled only as far and as fast as their feet could carry them. They didn't drink from plastic water bottles or spew pollutants from tailpipes. They had no electronic screens or twenty-four-hour fitness centers or microwaves to heat post-workout late-night burritos. And they certainly never drew arbitrary lines on a map to dictate the time.

Losing track of our internal time has proven disastrous. It's fair to say our clocks have gone cuckoo. The crazy thing is, we could have predicted our predicament. Humans have long been compelled to measure the passage of time with sundials, grandfather clocks, and neon-colored Swatches. And, like all species, humans are instinctively attuned to the sun's twenty-four-hour cycle. We seek homes with south-facing windows and skylights and carry babies outside for doses of sunshine. We buy blackout curtains. We may notice how much sharper and more energetic we feel at certain times of the day and lament the out-of-whack feeling that inevitably comes after a long-distance flight, or even with certain phases of the moon. Now, the science is finally catching up with our shared intuition.

While references to living rhythms can be found in the Bible and ancient Greek and Chinese texts, modern science has only recently exposed our internal clock's elaborate machinery, the astounding breadth of its functions, and its profound sway over our bodies and our culture. Only in the last few years have we begun connecting the dots and developing the tools to put this information to life- and world-changing use.

Armed with advances in biology and technology, today's renaissance in circadian science is uncovering potent strategies to recover lost rhythms and to leverage their power for a healthier, happier, and more sustainable and equitable world. Leading science and innovation organizations are joining in. In the US, the Defense Advanced Research Projects Agency, or DARPA, is developing implantable and ingestible bioelectronic devices to manipulate circadian rhythms, while NASA is enlisting special sun-mimicking lights to keep astronauts' bodies connected to the twenty-four-hour Earth day.

These efforts couldn't come soon enough. We can and need to make many more moves, as both individuals and a society, to reverse the escalating epidemic of broken circadian clocks. The list of things under circadian control is so long and the driving factors so ubiquitous that you'd be hard-pressed to find anyone unaffected. Yet many of us remain unaware that we've been living and performing suboptimally. Do you struggle to get a good night's sleep? Are you among the approximately one in five employed Americans who work outside of traditional daytime hours? Are you among the nine in ten who work indoors? How about the 99 percent of us affected by light pollution?

The groundbreaking discoveries in the burgeoning field include a link between light exposure and the healthy development of the brain, even while a fetus is in the womb. There is research that shows the coordination of our circadian clocks affects the composition of microbial squatters in our guts, and vice versa. There are also studies that show that restricting meals to daylight hours—the shorter the window, the better—could help stave off cancer, diabetes, and a host of other diseases. Even adjusting the time chemotherapy is administered may bolster its ability to attack some tumors. The COVID-19 pandemic brought further attention to the importance of biological timing. Our immune system is under the control of our clocks, and keeping clocks strong and well tuned is critical for fighting off contagions. SARS-CoV-2 is no exception. COVID-19 data further showed that the time of day a person is exposed, tested, or vaccinated may affect their prognosis, results, or level of protection.

One scientist I spoke to called the pandemic a "huge chronobiological experiment." Many of us revised our daily routines while setting personal records for screen time. We may have noticed that these changes affected our sleep, mood, and productivity. Some of us enjoyed an uptick in our sleeping hours; others endured more sleepless nights. Whether we were aware or not, lots of us learned how to listen to our biological clocks, not just the ticking of the clock on the wall.

Additional natural experiments emerged during this unique time and provide further circadian lessons, including a professional basketball league that finished the season in a "bubble" and a revised winter mission for the *Polar Star.*

The Coast Guard did not implement any onboard measures to compensate for the diminished circadian prompts awaiting the icebreaker in the Arctic that winter. Dim fluorescent lights lit the cramped hallways as crew members performed round-the-clock shift work. Captain Woityra reported to me from the ship in late December as it punched through two-to-four-feet thick sea ice north of the tip of Alaska. "On the way up, we had been experiencing a few hours of dusk in the midafternoon, but by the time we got there, it was full dark, all day," he wrote. After spending the weekend feeling lethargic, lacking motivation, struggling to get out of bed, and overeating, Woityra told me he pulled out the "happy light" for the first time on Monday morning: "It is a bit of a shock! I haven't seen this much light since I left Seattle!"

Captain Woityra was incessantly teased but kept up his daily doses of photons, basking in the lamp's glow as he caught up on morning emails, news, and the occasional crossword puzzle. He also discovered and began to use his computer's Night Light feature to filter the screen's blue light at the end of the day. Within seventy-two hours of starting his new light routine, he noticed major improvements in his sleep, his mood, and his eating patterns. "Everything just started to snap back," he told me. And he wasn't the only one seeing the light. A couple more sunlamps made it onto the ship, albeit as white elephant gifts—a fine alternative, I think, to the leg lamp from *A Christmas Story*.

I hadn't forgotten myself in my December shopping spree. I purchased a sun-mimicking lamp of my own as well as a quarter-sized clip-on sensor to measure my light exposure and a smartwatch to track my movements and sleep. I also began to contemplate taking my own journey into the "full dark," to discern my own rhythms and see what tweaks I might make in my daily life to optimize them.

As I learned more about circadian rhythms, I couldn't help but wonder if any of my lifelong sleep and seasonal struggles, even stomach pains, might be tied to my inner clocks. I've known wired nights, followed by dazed days, throughout my life. I've been almost guaranteed to suffer insomnia before significant events: basketball games, important tests, big presentations. I've attempted to compensate for sleep deprivation with copious cups of coffee and late mornings in bed on the weekends. My problems peaked in my teens. This is when our clocks are naturally programmed to shift later. My high school's 7:30 a.m. start time already felt unbearably early, yet I had to set my alarm every day to 5:30 a.m. so I could commute with my dad, a science teacher at the school. I'd also felt the effects of the rapidly expanding and contracting daylight hours across the year in Seattle, which I've only recently recognized is well north of the Canadian cities of Toronto and Montreal. And let's not forget that Seattle is pretty consistent in applying a thick gray sunblock. My years in Massachusetts and New York weren't much brighter. The seasonal patterns have always altered my mood, dampening it through the winter months. Yet it wasn't until January 2016, when I ventured down the dugout tunnel into the windowless locker room of the Seattle Mariners, my hometown's Major League Baseball team, that I began to see how my cycles of internal woes connected to cycles in my external world.

On a whim, I had taken up a public relations offer for a nighttime tour of the first LED-lit MLB stadium. It started with a dazzling show of the new field lighting. Our tour group watched the LEDs light up the grass as if it were noon and then put on a bright, colorful dance as if celebrating a Mariners home run. While less flashy, what I next saw inside captivated me more. The novel lighting system in the locker room allowed the team to control the intensity and color of the light fixtures. As a guide explained, the installation aimed to energize players before the game and relax them after so they could sleep and recharge for another nine innings—usually scheduled for the next day. I left captivated and a bit skeptical, just as most Mariners fans begin each new season. At the time, the Mariners hadn't made the playoffs in fifteen years. How desperate was the team getting? Did they really think they could start winning with the flip of a switch?

The experience flipped a switch for me, regardless. Circadian connections began to pop up everywhere. I soon realized I was watching a nascent field launch a scientific revolution.

As I saw circadian science proliferate, I noticed a parallel explosion in prospective applications. My curiosity was piqued: How could taking a walk in the morning and avoiding food after sunset keep your body clocks in sync? How might mimicking daylight indoors with fancy new lighting systems do the same? Why are some doctors prescribing treatments at specific times of day? Why are some leaders urging consideration of circadian rhythms to reduce economic and health disparities? And how might manipulating circadian clocks help curb climate change, reduce pesticide use, and ease world hunger?

As I began to unpack these and other questions, I encountered surprising historical and cultural connections—from a Georgian-era window tax in Great Britain to controversies around ending daylight saving time and delaying school start times. I followed a long tradition of circadian explorers and slept in a Cold War–era bunker, a sunflower field, and under the midnight sun of Alaska. Along the way, I spit into test tubes and plucked hairs from my head in efforts to decipher my own clock time, and I met many scientists and students, astronauts and athletes, and patients and policymakers at the forefront of a movement that may have found its time.

It now seems hard to ignore the modern menaces to our clocks and their staggering toll. Thankfully, many promising solutions are emerging. Some moves are surprisingly simple, like labeling when a bottle of breast milk was pumped or moving a work desk a few inches closer to a window. Living in greater harmony with our circadian rhythms might not solve all our ills, but it could go far in helping us live healthier, wealthier, and wiser. We're on the crest of a wave of potentially radical change. How far it carries us individually and as a society depends largely on how open we are to reconsidering our relationships with nature and time.

Part I

CLOCKS

1

Losing Time

With each step I took down the descending slabs of lumber, my surroundings grew a little cooler and dimmer. I welcomed the chilly air after the baking Arkansas heat above. Daylight, however, I was less thrilled about losing. I squinted up several times during my plunge, checking in on the shrinking patch of sunlit sky framed by the concrete stairwell. By the time I reached the bottom of the access portal's four and a half flights, the sky was merely a miniature rectangle far above. A few steps farther and I stood facing a foot-thick steel blue door. "You're gonna just pull hard, slow, and steady," my host instructed as I reached for the metal handle. "This weighs about the same as my truck."

The door was as heavy as advertised. I muscled it open, taking a full eighteen seconds. To the right, I could now see the long barren tunnel leading to a silo that once held a Titan II intercontinental ballistic missile with a nine-megaton nuclear warhead. To my left was another six-thousand-pound steel barrier. Slowly and steadily, I powered through that door, too, revealing a shorter tunnel that led to a former launch control center. This deep, windowless space would be my home for the next ten nights.

I'd come across GT Hill's vacation rental listing while scouring the internet for a place where I could fully hide from the sun, clocks, and people. Totally normal, I know. Perhaps, like me, your life has revolved around these things. I'm almost always seeking sunshine, racing against time (deadlines, usually), or frolicking with friends. But now, to better understand my body's internal rhythms, I wanted to avoid anything that could tip me off to the time. That included cycles of light and dark, and the clocks glowing on everything from laptops to microwave ovens. While the ability to survive a nuclear attack wasn't necessarily high on my list, I figured it couldn't hurt. And Hill's "Superhost" badge on Airbnb added an appreciated sense of security.

Still, friends and family in Seattle weren't shy about their concerns. Some questioned the sanity of my plan; others worried whether I'd have any sanity left by the end. I assured them, and less effectively convinced myself, that I would be okay. Hill promised to help.

HILL, AN AIR FORCE VETERAN WHO NOW WORKED REMOTELY FOR A Silicon Valley tech company, admitted to being a "bit crazy" himself. He spent more than half a million dollars and a decade's worth of spare time remodeling the decommissioned nuclear missile complex with its three-story, 3,500-square-foot launch control center. There were a few close calls along the way—like when he first pried open the flooded facility from the bucket of an excavator and 250 tons of water nearly washed him away.

Between 1962 and 1986, this center was filled with desks, switchboards, and everything else necessary to maintain one of the largest warheads the US had ever operated. That included a pair of keyholes and keys—kept in a safe—that could launch the missile. To strengthen the fail-safe, two people would have had to turn the two keys within two seconds of each other. The space also housed basic eating and sleeping quarters to support crews during their twenty-four-hour shifts. Today, thanks to Hill, the space flaunts a few upgrades: a 144-inch projection screen TV, a sprawling kitchen with a full-size freezer and flat-top stove, two refrigerators, two Keurig coffee makers, a floating king-size bed, and a bidet. I wasn't exactly going to be roughing it.

Hill and his wife had picked me up in Little Rock in his black Ford truck earlier that day. We drove the forty-five minutes north to Vilonia, down a series of increasingly narrow roads through the Arkansas countryside—over rolling green hills, alongside cow pastures, and finally down the gravelly Missile Base Road. You wouldn't have known anything special was here from a Google Maps satellite view. An American flag and a black-and-white TITAN RANCH sign spanning the side of a storage shed were all that marked the presence of the bunker. That, and a dark rectangular hole in the ground.

The ubiquity of man-made clocks is similarly easy to overlook, until you want to avoid looking at them. They pervade our lives just as molecular timepieces proliferate in our bodies. While preparing for my bunker trip, I'd spent hours trying to program the displayed time off the electronics I would be taking—a laptop, iPhone, Kindle—to no avail. I resorted to black electrical tape. Meanwhile, Hill prepped for my arrival with his own roll of tape. He covered the times displayed on the three tablets I would use to control the lights, and the digital displays on the microwave and the oven. As he toured me through the bunker, we caught and taped over two more digital clocks on the washer and dryer.

In the weeks leading up to my arrival, Hill and I had also devised a suite of communication strategies, including voice messages I would leave via a landline phone, written notes slid into and out of the bunker, and signals relayed via digitally controlled lights on all three floors. He wanted to ensure both that I was safe and that he never gave me any clock clues. Just one hint or glimpse of the time during my ten days underground could ruin the experiment.

Then there were the bonus gadgets I enlisted for the experiment, some also brandishing the time, now under electrical tape. In addition to taking notes on my daily routine—as well as my mood, hunger, alertness, coordination, and cognition—I planned to collect data on as many physiological rhythms in my body as possible during my stay underground. Around both wrists and one finger, I wore a sampling of wearables to measure things like heart rate, activity, and sleep. I clipped a light sensor onto my shirt and attached a continuous glucose monitor to my stomach, a tiny needle penetrating my skin. I also taped tiny temperature sensors on various body parts.

Benjamin Smarr, a data scientist at the University of California, San Diego, lent me those temperature readers, each about the size of a coin cell battery, and offered to help me analyze the information they and my other devices would collect. In the same yellow package as the stack of six sensors that he mailed me was a handful of nitrile gloves. Just in case I was up for it, he told me, I could cut the fingers off those gloves and use them to place one of those sensors more, let's just say, "internally." I opted against that after rational deliberation about safety and comfort.

Without access to the sun or other cyclical cues around us to calibrate our clocks, the internal timekeepers scattered inside our bodies will keep ticking on their own. But they are unlikely to measure a day as precisely twenty-four hours long. A "day" in the life of our body will instead be "about a day," or *circadian*. In Latin, *circa* means "about" and *dies* means "day." Once untethered from reliable daily signals, our clocks will begin to drift, carrying along various cycles in our physiology, from sleepiness to strength. What's more, our clocks can slip out of sync from one another. The clocks ticking together within the liver may no longer keep the same rhythm as those in the skin, nose, or heart. If you've ever had jet lag, you've experienced what happens when body parts start beating at different tempos. The contradictory cues can cause a range of unpleasant effects, such as headaches, digestive troubles, and difficulties concentrating.

Our body clocks rely on one another. "They want to be synchronized," said Smarr. "They want to know that everyone is where they are supposed to be. But they can't see each other very well." If you're a soccer fan, you could imagine how a left midfielder who can predict the whereabouts of a forward on the right side of the pitch in four seconds—approaching far post, for example—will be more apt to place the ball at their feet and increase their team's chances of scoring a goal. "If they're not there, your team is not very effective," said Smarr. And so it goes for our health, too.

Ultimately, it may not just be that my bunker days run long or short. My whole team of clocks may fall apart. That could result in my temperature and glucose rhythms no longer aligning with each other or with my sleep or heart rate. Life in my body may become a "fractured experience," warned Smarr: "Who knows, maybe ten days gives you enough time to go crazy for a day or two?" Even so, I would be lucky. Whatever mess would become of my rhythms should be temporary. I am fortunate enough not to be a shift worker nor still a high school student forced to wake up hours before my body's rhythms are ready. I also live in an apartment with decent daylight and without bright streetlights beaming into the bedroom.

Before we hung up, Smarr reminded me to keep track of when I go to the bathroom—yet something else driven by circadian rhythms. It was becoming increasingly clear to me that nearly everything abides by the inner clock, and that I might need to recalibrate what I considered TMI.

THE SUN, CLOCKS, AND PEOPLE WOULDN'T BE THE ONLY THINGS I cut myself off from in the bunker. I had recently learned that the color least likely to impact human circadian clocks is red. So I programmed Hill's digitally controlled lights to a dim red, bright enough that I wouldn't trip down the stairs but muted enough that the lights shouldn't mess with my sleep. I supplemented with a red headlamp, orange-hued reading light, and electric candle. But as I unpacked my stuff, I discovered a loss I hadn't anticipated: nature's rainbow of colors. I now saw only shades of reds and pinks, browns and grays. All other color vanished, including the eggshell blue of the typewriter I had borrowed for the experiment.

After organizing my luggage and putting away my groceries, and after making my first use of the upscale bathroom, noting the clock time, I placed a piece of now pinkish paper into the typewriter and began drafting a few words of this chapter. Tapping the stiff keys made a satisfying clicking noise, which echoed off the concrete beams and walls. Adding to the mechanical melody was the typewriter's pleasant "ding" that warned me I was about to hit the end of the line and should pull the carriage return to start a new one. This happened too soon with one line—a common malfunction with the aging 1970s Kmart 300 Deluxe 12 typewriter.

A couple of rough paragraphs later, I glanced at the time on my pink Fitbit, the one clock I had temporarily left visible. It was getting late. At 11:15 p.m., I covered this last digital screen with black tape and declared myself time blind. The experiment began. Many questions ran through my mind as I walked up the two flights of spiraling steel stairs to get ready for bed: How would I know when to call it a day the next night, and the nights after that? Would it even be night? There would be no watch on my wrist or anything else external to warn me that the day was almost through. I would be wholly reliant on my internal timekeeping system. Would those clocks protect my days from escaping the boundaries they had known for the past forty-plus years, or would they leave me transitioning to a new day too late or too soon—like the broken right margin stop on that fortysomething typewriter?

I WOKE UP GROGGY ON MY FIRST MORNING IN THE BUNKER. IT'S A FAmiliar state for me. But this was the first time I couldn't confirm by watch or phone how much, or how little, I'd slept. And the fact that it was still dark, not even a sliver of light peeking around curtains, didn't eliminate the possibility that the sun was up.

After lying awake for another unmeasurable length of time, I finally got up and stumbled down the stairs to the kitchen. There, I initiated what became my morning ritual of reading over breakfast and Keurig coffee—Clover Valley's Donut Shop Blend medium roast with a shot of oat milk, in a white Titan Ranch mug. I also made my first guess of the time into my voice recorder: 9:10 a.m. The time stamp would later reveal that it was 9:39 a.m. Not bad.

For the record, my closest guess during the ten days would be just nine minutes off. That came during my second full day, when I left one of my regular check-in voice messages to Hill. I reported to him that it was about 1:00 p.m. As he later told me, my call came in at 1:09 p.m.

My sense of time would not stay so accurate. Still, at least for the first few days, I remained curious about what the clocks around me read under the tape, and how my internal tickers were coping with the unchanging scene: the constant dimness, the static temperature and humidity, the lack of fluctuating breezes and shifting colors. After my third night, I labeled my diary entry "day after 3rd sleep" and stuck to that nomenclature for the rest of my stay. My then boyfriend had opted for the term "phase" when numbering each of the sealed notes he gave me to open underground. There were fourteen of them, just in case my clock ran short.

I filled my time with reading, writing, flinging juggling balls, and puffing into a harmonica. I couldn't think of a better place to take up the latter hobby because no one would have to suffer the sound. After three or so days' practice, I grew brave enough to perform in front of the security camera I had set up for one-way communication with my family and friends. I had been working on basic juggling tricks—behind the back, under the leg, the "shower"—and a simple harmonica version of "When the Saints Go Marching In." The pressure was low. I couldn't see the reaction of anyone watching my live feed. I would perform a few more times during my stay, eventually adding "What a Wonderful World" to my repertoire. Only later, during a phone call at the airport on my trip home, did I get feedback—my then boyfriend pointing out how variable my skills appeared depending on the time of day. The ease with which I kept balls in the air dropped precipitously as the evening wore on—at least during the early part of my experiment when his and my days and nights more closely aligned.

My focus and brainpower also varied across the day, as did my mood and anxiety levels. Within an hour or two of getting up each day, according to reports on my voice recorder, I would predictably feel various combinations of "happy," "motivated," "optimistic," and "excited." There were other hours—usually in the middle of my "afternoon"—when I felt lower and lonelier and often had to read the same paragraph of a book multiple times. I couldn't fault dense or boring material either. My first bunker read was Mary Roach's *Fuzz*, about the places where humans and wildlife meet or, more accurately, come into conflict.

AT THE ENTRANCE TO THE KITCHEN, UP A HALF FLIGHT OF STAIRS from the control center's original exit hatch, a pair of flyswatters hung on a steel cabinet. Both bore a camouflage print that appeared orange and brown under the red lights.

My third "morning" is when I first realized these swatters' utility. I was not alone down there. The occasional tiny fly began to dart around my head as I ate, as I read, and as I wrote. They did not seem to stick around as I practiced the harmonica. But who could blame them? Still, I never put the swatters to use. Compared with their blackfly cousins, fruit flies are exquisitely polite. I soon caught myself enjoying the regular company of what I convinced myself was one friendly fruit fly. It had a quiet, calming presence. I never noticed a buzz. And I named my new friend Per. Sure, this may have legitimized my friends' concerns. But bear with me.

It was September 2021. Exactly fifty years had passed since a landmark moment in circadian science. By the late 1960s, hints of circadian rhythms being endogenous—meaning, driven internally and not by the outside world—spilled from across the animal kingdom. But no one knew where exactly the clockwork of life resided nor how it worked. A breakthrough by scientists Seymour Benzer and Ronald Konopka at the California Institute of Technology began to zero in on that timekeeping machinery. The scientists bred and reared thousands of tiny fruit flies in bottles. After exposing the flies to a poison that triggered genetic mutations, Konopka, a graduate student in Benzer's lab, closely watched the flies' resulting behavior. He found three strains of mutant flies with a warped sense of time: one strain lost all rhythmicity, another strain's days shortened to around nineteen hours, and the third's days lengthened to approximately twenty-eight hours. Even more curiously, all three mutations mapped to the same gene in the fly. Somehow, that gene programmed the period of a fly day. So, naturally, Konopka and Benzer named that gene *period*—or *per*, for short. They published their findings in September 1971.

Konopka had "stumbled straight into the center of the living clock," writes Jonathan Weiner in his book *Time, Love, Memory*, which chronicles research into the genes that regulate these three feats in fruit flies. The book followed *Fuzz* in my bunker reading. A few decades after that fortuitous stumble, a trio of circadian scientists who built on the work of Benzer and Konopka and others won the Nobel Prize for decoding many of the molecular mechanisms controlling circadian rhythms. Included in their findings were two more clock genes that coordinated with *period*: *timeless* and *double-time*. Other researchers have identified a number of additional components of the fly's molecular clockwork as well as of our own. Despite our differences, we share very comparable clock genes with flies. Decades of discoveries mounted to a remarkable revelation: evolution bestowed early organisms with internal pacemakers, and their impressive survival benefits kept these microscopic cogs and gears in play. "Inventing a clock was probably one of the first acts of life," Weiner writes.

Reading Weiner's book in the bunker, I came upon a seemingly unfortunate piece of fruit fly trivia: the insect's scientific name, *Drosophila*, means "lover of dew." In other words, fruit flies are morning creatures. Worrying thoughts immediately filled my head: If Per was most active in the morning, then was its presence a clue to the time of day? I had worked so hard to avoid seeing daylight or clocks; now, I feared my newfound companionship could spoil it for me. I eventually rationalized that Per had probably spent its whole life in the bunker, which meant its circadian clocks had never tethered to the sun. We could keep hanging out, I decided. Thankfully, it didn't occur to me until weeks later that Per had more likely hitched a ride down to the bunker with my bananas or mandarin oranges.

MIDWAY THROUGH MY UNDERGROUND STAY, MY CLOCKS WENT HAYwire. Smarr and I later determined from my data that various rhythms in my body started to fall out of sync the day after my fifth sleep, just as he had predicted might happen. My voice recordings subsequently mention a "distant and grumbly" stomach, a "sunken feeling," a "heavy chest," and feeling "loopy," "woozy," and "weak." I describe being hot and cold at odd times of day. And I've since deemed one of those days the clumsiest of my life: I managed to drop not only juggling balls but also my voice recorder, Kindle, and harmonica. While the symptoms were more extreme, they felt familiar; they reminded me of my body's disheveled state during the first several days of a recent vacation in Vietnam. It felt like jet lag.

I was certainly not the first person to experience jumbled clocks. I was also far from the first to hide from the sun and society for the sake of science. A long tradition of circadian explorers preceded me. On June 4, 1938, Nathaniel Kleitman and Bruce Richardson, then researchers at the University of Chicago, went deep into Mammoth Cave in Kentucky. A small cavern 119 feet belowground became their home for thirty-two days. Isolated from any fluctuations in light, temperature, or sound, they read by lantern and ate fried chicken, country ham, and other meals prepared at the Mammoth Cave Hotel. They devised traps to keep rats out of their beds.

The researchers also brought down an alarm clock. They would use it to break the thirty-two calendar days into about twenty-seven strictly scheduled twenty-eight-hour "days," and see if the human body could adapt. In the end, the younger researcher, Richardson, succeeded in following that twenty-eight-hour day. Kleitman, on the other hand, seemed stuck on approximately twenty-four-hour cycles. It was a curious result yet did not inspire much initial interest. Decades later, that would change. Scientists eventually appreciated how well Kleitman's sleep-wake behavior and body temperature nearly matched a twenty-four-hour cycle despite the twenty-eight-hour time cue and the complete lack of daylight or other earthly signals. "What he had actually discovered was the circadian free-running clock. He just didn't know it yet," Russell Van Gelder, an ophthalmologist at the University of Washington in Seattle with expertise in circadian rhythms, told me.

Michel Siffre of France began a few related experiments more than twenty years later. On July 16, 1962, the twenty-three-year-old went 375 feet deep into a glacier in the French Alps near Nice. Conditions were rough, to say the least: below-freezing temperatures, 98 percent humidity, and falling rocks and ice. And he was utterly alone, minus the spider he would trap and keep in a box as a pet. I did feel slightly less self-conscious about my fruit fly friendship after reading of his attempts at companionship. Although I probably should've felt more embarrassed about my three-story climate-controlled accommodations with a floating bed, faux fireplace, and four showerheads. Siffre's living space consisted of just 115 square feet. That was enough for a chair, a table, a stand of shelves, and a camp bed—on the head of which he clamped one weak light bulb.

When Siffre emerged after sixty-three days, he thought little more than a month had passed. You might say he was overthinking it. On his fifty-first awakening, Siffre made a note estimating that he had spent fifty-two days in the cave. Yet he believed his daily cycles were only fourteen hours long. So he translated that to approximately thirty-two days of twenty-four hours each and mourned the "thought of a whole month more still to go." While his brain's perception of the passing time might have been off, his body kept a nearly normal daily beat of sleeping and waking. Siffre's data would later reveal that an average day in his body was twenty-four hours and thirty-one minutes. That data added support for the presence of circadian rhythms in humans, and it provided more evidence that the cycle of those internal rhythms may be close to, yet not exactly, twenty-four hours long.

Around the same time that Siffre was spelunking, a series of experiments got underway in a German bunker constructed deep in the Bavarian countryside. Jürgen Aschoff, a pioneering circadian scientist with the Max Planck Institutes, had been studying the inner workings of biological clocks in plants and animals, including birds, hamsters, and flies. Rhythms appeared everywhere he looked. But there was one animal he had yet to investigate. "My co-workers and I became eager to know whether man also possesses a circadian clock," Aschoff wrote.

Aschoff and his colleagues initially began their human experiments in a deep cellar of a Munich hospital. Still, to really tease apart what makes us tick, Aschoff determined that they would need a space void of any hints of time from the outside world—radio, television, clocks, phones, or fluctuating environmental clues, such as temperature, daylight, noise, or vibrations. Rather than seek out a cave, they built one: a windowless chamber with meter-thick reinforced steel walls in a hillside just down the road from Kloster Andechs, a monastery famous for its craft beer. Then they managed to persuade a steady stream of people to take up temporary residence in the two small bunker apartments. Many were students seeking a distraction-free study space.

Aschoff's team asked participants to lead a regular life inside the time-free apartments, including cooking themselves three meals a day and going to bed when they were tired. In addition to fresh food and other necessities, participants received one bottle of Andechs beer daily. They more than earned those beers. Being part of the study meant using a rectal thermometer and providing regular urine samples.

On the last day of a participant's stay in the bunker, the scientists left a note saying they would be stopping by for a visit. They did not divulge the reason. When the scientists entered, they asked the participant to guess the time and day of the week. Everyone got it wrong. In the end, an average "day" for 85 percent of the participants exceeded twenty-four hours. At the extreme was one man who had spent five weeks in the bunker despite thinking it had been only three. His sleep-wake cycle was even more impressively off at about fifty hours. As he left the bunker, he lamented the weeks he lost.

The scientists discovered something else peculiar: For many participants, the timing of sleep and activity, and daily fluctuations in temperature, urine excretion, and other body functions, lost their alignment. The daily up and down of temperature might have followed a cycle of around twenty-five hours, for example, while the rhythm of sleep and wakefulness recurred every thirty-three hours. This made them wonder whether there might be *multiple* clocks ticking inside us.

Charles Czeisler, a sleep and circadian researcher at Brigham and Women's Hospital and Harvard Medical School in Boston, is among scientists who have since designed and run protocols that collect broader arrays of health measures with more stringent restrictions on sleep, activity, eating, even body posture. The aim is to disentangle true circadian effects from the direct impacts of behavior and the environment on daily rhythms. In some of these intensive experiments, he's imposed eleven-hour, twenty-eight-hour, even nearly forty-three-hour days on participants. Because it is impossible for a person's circadian system to synchronize with schedules so far from the twenty-four day, the idea was that participants' clocks should revert to their internally produced, endogenous circadian period. The work led Czeisler to calculate and fine-tune what is widely considered the most accurate measure of the average healthy human daily rhythm: twenty-four hours and nine minutes.

He finds circadian cycles in sighted people rarely run shorter than twenty-three and a half hours or longer than twenty-four and a half hours—much tighter than the ranges of rhythms calculated in the Bavarian bunker, where access to electric lights probably skewed participants' circadian periods. In general, people whose clocks run significantly shorter than twenty-four hours tend to be morning larks. Those whose clocks run significantly longer are likely night owls. The data suggest there are more owls than larks among us, with most people falling somewhere in between.

But maybe the most important revelation from all the experiments to date is the natural range of rhythms between people. Today, Smarr and others are bringing attention to an overlooked form of inequity permeating society. Traditional school and work schedules favor people with faster clocks, making life harder for owls and their slower clocks. "We are discriminating on someone's biology," Smarr told me during one of our pre-bunker conversations. "It's not something we can see, but it's not okay."

BY THE NIGHT AFTER MY EIGHTH SLEEP, I'D ALREADY CONSUMED MY choice dinner options—the cans of chili, the frozen pad Thai and curry dishes. But I did still have the fixings for pancakes. Breakfast for dinner has been a lifelong favorite of mine anyway. So I mixed up some batter, spooned it onto a frying pan, and plopped in an abundance of frozen blueberries—which looked more black than blue under the red lights. Before I took my first bite, I recorded my guess of the time: 5:00 p.m.

Not even close. I began eating just after 6:00 a.m. It was breakfast time. At that point in my experiment, I had unwittingly reversed my sleep-wake rhythm. I was now working the night shift, just as many Cold War–era crew members had done while manning this center decades before and just as millions of Americans do today.

I wasn't naive, of course. I never assumed accuracy with my time estimates. I could feel the chaos in my body. And the underground experience wasn't only reminiscent of flying internationally. I've felt similarly off at other times in my usual aboveground life. I would go on to realize that many of us live with whacked-out rhythms every day due to our indoor, 24-7, high-tech world. By inventing our own version of time, we have loosened our link to the natural rhythms of the Earth and the sun. The results are threatening real consequences for individuals and society.

I would also learn that this doesn't have to be our reality, nor is it currently the reality for everyone. As I took my last bites of the blueberry pancake, I thought of James Copeland. Little did I know, as I finished dinner and began winding down for what I thought was my second-to-last night, he was just waking up—a few dozen feet above me in his Winnebago.

I had met Copeland upon my arrival at Titan Ranch. He lived and worked out of his wheeled home parked on the property. When he wasn't helping Hill run the facility, the twentysomething wrote radio scripts on his many typewriters, one of which he generously lent me. The green-striped Winnebago and nearly everything I saw inside it, including an eight-inch black-and-white RCA tube TV, dated back forty-plus years. The black Converse All Stars on his feet were a fitting touch.

In line with his old-fashioned disposition and style, Copeland seemed to live more in tune with the sun's cycles than most of us in the modern world. He ate three meals at about the same time each day. He claimed a regular sleeping schedule: to bed between 10:00 p.m. and 11:00 p.m., and up by around 7:00 a.m. every day. He used blackout curtains sewn by his mom to keep the Winnebago dark at night. Plenty of light streamed in through those same windows during the day. The small square footage meant he was never far from a window. Plus, he was often outside during those hours anyway—at least when he wasn't underground. In the evenings, Copeland wound down by listening to shortwave radio and reading. He rarely struggled to fall asleep. In many ways, he was the model of circadian living, a time capsule into a former, more sun-synced way of life. Where have we gone wrong?

I spotted a few clues in the bunker. Some of Hill's modern touches reflected aspects of humanity's assaults on our circadian clocks. The refrigerator situation was one example. Hill usually kept the kitchen fridge well stocked with soda for his guests. For my stay, he removed some cans of Coca-Cola and Mountain Dew to make room for my ten days' worth of provisions. There was an upstairs fridge, too, beside the bed. The Western world is obsessed with sugar, caffeine, and round-the-clock eating. The modern diet, both what we eat and when, might be nearly on par with artificial light in messing with the hands of our clocks. And these insults can go hand in hand. Opening a fridge at night provides a dose of potentially sleep-derailing light. I knew enough at this point to sidestep that obstacle during my stay: I attempted to memorize the general location of everything I put in the fridge. That way, I could close my eyes and feel my way around. Mandarin oranges? Third shelf down, to the left of the eggs. Then there were the fancy lights that glowed incessantly above each mirror in the open bathroom, visible from the bed. I eventually draped towels over both.

Yet Hill's design work, unbeknownst to him, also showcased emerging innovations that I was learning could help rescue our rhythms. While they won't send us back in time to live exactly like our ancestors, or even necessarily like Copeland, they could help us realign our clocks closer with one another and with the sun. Take Hill's tunable lights. Both the intensity and color of light—and when and how long we look at that light—influence how those photons affect our circadian rhythms. If we can thoughtfully harness those factors, which systems like his allow us to do, then we can better tune our clocks. Further, Hill decided against placing a glowing digital alarm clock beside the bed as traditional hotels are still known to do. The intrusive digital lights can disrupt our sleep. And use of the alarm clock itself can keep internal clocks from orchestrating optimal rhythms. Still, Hill did help me devise an alarm of sorts for the one morning I would need it.

Thinking about that imminent alarm caused me anxiety my entire last day underground. Unless my internal days were unusually short, I knew my bunker time must be nearing the end. My body had recalibrated slightly from the previous day, and I finished dinner by 2:30 a.m. Even so, I felt exhausted after the meal. I finally gave in and went to bed shortly after 4:30 a.m., guessing into the voice recorder that it was about 8:00 p.m.

Two and a half hours later, as planned, Hill digitally flipped on all the lights in the bunker from his home fifty-some yards down the gravel road. My eyes shot open. The jarring onslaught of photons also sent my heart pounding.

Much like on that first morning, I lay in bed bleary-eyed and feeling as if I had only just fallen asleep. This time I could substantiate my suspicion. I reached for my iPhone, which I had kept powered on and near during the experiment to ensure my experimental data uploaded to the cloud. I was about to take my first look at its screen in ten days. Peeling off the black tape, I confirmed the time. I also checked the time stamps on my voice recorder. Sure enough, based on my bedtime recording, I'd slept about two hours and fifteen minutes.

I then began stripping black tape from all my devices and flipping on the remaining digital light switches throughout the bunker—I went for full blast and full spectrum. Light and color streamed back into my world. So did the clocks, along with the texts, emails, and social media notifications. Next came my reunion with humans.

Hill and his family met me in the bunker kitchen to debrief. One of Hill's first questions: "Do you know what a Fibonacci sequence is?" He spoke of a mathematical pattern in which you add the previous two numbers to get the next number. Not exactly the conversation I was anticipating, yet it paired well with the rest of my abrupt transition back to reality. I let him continue. The differences between my estimates on phone calls and the actual clock times were "almost Fibonacci," he said.

"You were like, three hours off, four hours off, seven hours off, and then six hours off, but then thirteen hours off. You were never ahead. You were always behind."

Soon, I muscled back through the steel doors. That small patch of bright light reappeared overhead. I squinted up at it, then quickly cast my eyes back down while shielding them with my left arm. My eyes started watering. I'm sure this was largely reflexive of eyes now accustomed to the dark, but I might attribute at least a couple of tears to my joy and relief at seeing daylight again. I put on my purple sunglasses and climbed the fifty-two stairs toward the bright, now enlarging rectangle. I would be reminded of this experience when I later read Siffre's account of surfacing from the French cave. He wore a massive pair of sunglasses with triple lenses, "very darkly tinted," over snow goggles.

My eyes could have seen worse. As much as I'd been dreaming about sunshine—and playing a harmonica song about "skies of blue"—I was grateful for that morning's cloudy, Seattle-like weather. I was also happy to learn from Hill that my Seattle Seahawks had won on Sunday. He, too, was a fan of the football team. The game was well over when I, via the security camera, had cheered them on in a message to family and friends back home.

The next few days were a blur. I had hours of giddiness—maybe because I was overly tired or simply savoring my reacquaintance with sunshine and people, breezes and colors. I also had moments of feeling very low and exhausted. My stomach remained a disaster for days. I vacillated even more between being unusually cold and unusually hot. I struggled to sleep at the appropriate time. My clocks, as predicted, were very confused. Now, it was time to take a closer look at their cogs and gears and understand why.

"YOUR DATA LOOK BEAUTIFUL." I BLUSHED WHEN I READ THE SUBject line of Smarr's email. Weeks had passed since I surfaced from the bunker. I had mailed Smarr back those temperature-sensing buttons, and he had run an initial analysis of my data. "Textbook," he called it. "You clearly drift then spring back when you emerge."

Smarr was one of several experts who helped me sort through the anarchy that had taken place in my body during and after my time underground. I met him and Ethan Buhr at the Society for Research on Biological Rhythms 2022 Biennial Conference on Florida's Amelia Island. The sunny locale seemed appropriate for the subject matter; the windowless conference rooms not so much. The three of us sat outside the hotel on lounge chairs, roasting in the afternoon sun and savoring every breeze that passed through the thick humid air. They had both been looking closely at my data, especially the measures of my body temperature.

As the body prepares for bedtime, skin temperature rises in parallel with a drop in core body temperature. This reflects heat leaving the body. A basic plot showed the pattern of my daily high and low skin temperatures over the ten days.

"You start so nice," said Buhr, a circadian scientist at the University of Washington.

"Thank you," I replied. I could get used to this data flattery.

But after the first four or five days, he added, "it was mayhem." Soon, the initially distinct crests and troughs grew less and less pronounced and drifted later and later in the day. Scientists define a circadian rhythm, or any oscillating aspect of our physiology, by its phase, period, and amplitude. The phase describes the timing of a peak, valley, or other specific point in the rhythm. The period depicts the time needed to complete a full cycle, such as from peak to peak, and the amplitude how high the peak climbs relative to the valley. As Smarr and Buhr explained, a greater amplitude generally marks a stronger, more robust circadian rhythm.

By day eight, however, my highs could barely be discerned from my lows—the opposite of a stellar rhythm. Even after I returned to Seattle and sprang back to diurnal, meaning I was again active during the day like most fellow humans, my daily temperature rhythms remained weak and muddled. It was a similar story with my heart rate. Daily highs and lows corresponded to when I was active or asleep for the first few days in the bunker. But then things fell apart. The data supported the discombobulation I'd felt in my body.

Smarr and Buhr also looked at the relationships between my temperature, heart rate, and glucose data—the times during the day that each descended and rebounded relative to one another. The periods of all my rhythms, including my sleep and wake pattern, ran longer than twenty-four hours. But the period of my temperature rhythms exceeded the period of my heart rate rhythm. They had uncoupled not only from the sun but from each other. Similarly, the initial coordination between the phases of my temperature and glucose rhythms quickly dissolved after about four days. "They start together and then end up wandering off on their own," as Smarr put it.

Individual parts of me were beating to their own drums. Smarr had likened the body's network of clocks to a soccer team; Buhr elaborated on the orchestra analogy. "Every member is totally capable of playing on their own. And they are really good musicians," he said. "But if all of a sudden they can't see each other, and they can't hear each other, and they can't see the conductor, their tempos would all get out of sync. It would sound like garbage." By around day six, my body clocks had lost their coordination and tempo. I felt like garbage. "But then if all of a sudden you reintroduced the conductor, and they could see the conductor, then they would drift back into sync," added Buhr. Thankfully, this, too, eventually happened for my clocks after about a week back in Seattle.

Olivia Walch, a Virginia-based applied mathematician and the CEO of Arcascope, which designs mobile apps based in circadian science, created a series of figures to help me visualize this inner chaos. The colorful images underscored the point, revealing similar disordered patterns and weakened amplitudes in my rhythms, whether looking at my temperature, heart rate, sleep, or activity based on step counts. On the final days of the experiment, my "days" and "nights" appeared almost indistinguishable. That is not a good thing, especially if the pattern is chronic. Studies now link low amplitudes with multiple measures of poor health and disease risks. Both synchrony and strength appear to be critical for healthy clocks.

It bears repeating that the circadian insults inherent in modern life can wreak similar havoc, perhaps even worse, than days spent underground. They can send our circadian clocks straying from the sun and from one another. Unwisely, we orient our lives counter to our clocks—counterclockwise, you might say. And we are only starting to learn the consequences.

2

What Makes You Tick?

Plant names sometimes sound silly, ill-fitting, or even offensive—humble plant, horehound, and sneezeweed come to mind. The sunflower, on the other hand, easily earns its cheery title. Its enormous round of yellow petals resembles rays of sunshine. Not to mention, it is famous for following the sun.

I drove twenty-five miles northeast from Seattle to the river valley town of Duvall shortly before sunset one warm July evening to spend quality time with sunflowers—hundreds of thousands of them. Lora Lee Wicks, of Lora Lee's A-Maze-Zing Sunflowers, stepped out of her brown farmhouse to greet me. Her husband and two sons followed, along with what at first looked like a puppy. "That's not a puppy, is it?" I said as they approached. It was in fact a two-week-old brown-and-white mottled goat named Daisy. Her mom had stopped feeding her, Wicks told me. So that job now fell on its human family. Wicks's teenage son, Dietrich, picked Daisy up into his arms and, with a huge grin, shared that he did the bottle feeding.

Wicks walked me around the back of her house, past a chicken coop, a large cage with four rescued ducklings (also abandoned by Mom), and rows of snapdragons, zinnias, dahlias, and sweet peas. We made our way down a narrow dirt road flanked by blackberry bushes and covered in knee-high sedges and mayweed, for which she apologized: "I'm good at growing weeds." Straight ahead, to the west, the sun was closing in on the crest of some hills. We high-stepped with our flip-flops down the overgrown path. It eventually opened into a grassy area backed by the main attraction: a massive field of stalky green sunflowers. The plants varied from about three to four feet tall. And they nearly all had their backs to us.

We snuck up behind several flowers to inspect their bulging green buds. They were about the size of golf balls. At the top of each, tiny tips of soon-to-be golden petals came together and pointed west to the hills. Occasionally, I noticed that a petal tip or two had begun splaying off. "They should bloom within the week," Wicks told me. Her website optimistically advertised that the season opener of her maze might be the coming weekend. This was only her family's second year in the maze business, and she was clearly excited. She boasted of the twisted paths that Dietrich and her husband, Ed, had created within the field—a maze in the shape of more sunflowers.

Wicks and I eventually wandered back to the goat pen so I could take a turn holding Daisy before setting up camp for the night. I pitched my tent on the grassy field abutting the backs of the sunflowers. For the next few hours, I watched the bright oranges and pinks of sunset turn into the deep violets and blues of dusk. Cows mooed in the distance. Birds squawked and shuttled in mass formations overhead, likely en route to join their nightly roosts. I was reminded of the hundreds of crows that gather daily in the park by my Seattle apartment during the fall and winter months. The cawing usually commences a couple of hours ahead of sunset before the murder then heads northeast about eight miles, as the crow flies, to a grand regional roost of more than sixteen thousand birds. I also thought of the plants and critters I had seen on and around Wicks's property: the weeds, dahlias, ducklings, Daisy, even the Wickses' real dog, Zuzu. Every one of them followed predictable daily patterns. And so did the sunflowers.

After the sky had fully darkened to reveal the stars, I took another look at the field of flowers. It was 10:30 p.m. Most of the sunflowers stood up straighter with their heads now tilted skyward as if gazing up at Hercules or the Little Dipper. It was the night before the new moon, and the sky was clear. I crawled inside my tent, leaving off the fly so I, too, could enjoy the star show. And I set my alarm for 3:00 a.m.

LIVING THINGS BEGAN TRACKING THE INCREMENTAL PASSAGE OF time long before the man-made clock lent its hands. As life grew in harmony with the sun's daily march through the sky, and with the seasons, phases of the moon, tides, and other predictable environmental cycles, evolution ingrained biology with the timekeeping tools to keep a step ahead. It gifted an ability to anticipate changes, rather than respond to them, and an internal nudge to do things when most advantageous and to avoid doing things when not so advantageous. Of course, that optimal timing depended on a species' niche on the twenty-four-hour clock. When mammals first arose, for example, they were nocturnal—most active during the hours that the cold-blooded, sun-needy dinosaurs slept. Now mammals occupy both their choice territories on a spinning planet and their preferred span on a rotating clock.

Timing is everything when it comes to seeking and digesting food, storing food, avoiding becoming food, dodging exposure to DNA-damaging ultraviolet radiation, and many more vital activities, such as navigating, migrating, and reproducing. Take the *Eudyptula minor*, a tiny penguin species that lives on Phillip Island in Australia. The slate-blue plumaged seabird speed waddles from the ocean to burrow home at the same "sun time" each day—just after sunset. Finding that precise window between day and night maximizes the penguins' fishing time, allows them enough light to see their way to their burrows, and minimizes the chances they become visible food along the way for nighttime predators, such as orcas, seabirds, and feral cats. A clock off by just ten minutes could prove fatal, one source told me. The island's tourism industry capitalizes on this predictable "penguin parade." A website lists approximate penguin arrival times for every day of the year and sells tickets to witness the spectacle. A higher ticket price grants visitors access to an underground viewing structure where they can watch the procession of waddlers at eye level. In October 2022, lucky visitors got to view a record-breaking 5,440 little penguins storm the shore and hurry home.

I doubt anyone would pay to see *Trypanosoma brucei*, the parasite that causes sleeping sickness. But its timely feats are extraordinary, too. The parasite, endemic in sub-Saharan Africa, leverages its time-tracking mechanism to carry out a life cycle that includes hitching rides on tsetse flies, traveling through the bloodstream of a human or other animal, and ultimately disrupting that host's circadian clocks. Patients sleep at strange times of day—hence, the name given to the deadly disease.

An arguably more charming example is the honeybee waggle dance. After returning to the hive from a successful foraging trip to, say, a sunflower, a worker bee will appear to waggle excitedly. A closer look shows she forms figure eights, shaking her abdomen as she moves across the middle line between the top and bottom rounds of the "eight." The duration of the waggling indicates the distance to the flower. And the orientation of this wiggly line in relation to the sun tells her hive mates the direction of the delectable nectar. But the dance includes a twist. Because that relationship to the sun changes hour by hour, the bee must continually update her dance, or she'll lead her mates astray. This meticulous choreography would not be achievable without a circadian clock—just as the flowering plant needs its biological rhythm to take full advantage of the bees' services.

For you and me, our clock network manifests as regular rhythms inside our bodies—such as the ebb and flow of hormones and rise and fall of blood pressure and heart rate—and in our behaviors. Clocks help our digestive and metabolic systems to gear up ahead of time to efficiently process a meal, assuming those meals arrive at similar times each day, and the skeletal muscles to fire at peak force when they are most needed.

Our strength generally maxes out around dusk. That's probably when our ancestral hunters would have hauled home their harvest. Most of us still organize our schedules somewhat similarly to those early humans, who restricted activities such as hunting and gathering to daytime, when they were less likely to fall off a cliff or end up a meal for a nocturnal predator. Around dusk is also when the circadian system instructs the pineal gland in the brain to begin releasing melatonin. This hormone tells the body that darkness has descended and, for us diurnal creatures, that it is time for rest. It doesn't directly make you sleepy but sets in motion other physiological processes that do. You can also likely thank your well-tuned clocks for slowing down your kidneys' urine production and enlarging the storage capacity of your bladder so you can sleep through the night without getting up to pee, and for spurring your adrenal glands to pump cortisol to rev you up for the new day. Your sleep, mood, appetite, immune response, sex drive, and body temperature all wax and wane under direction of your circadian rhythms. The list goes on.

Exposure to environmental cues such as the alternation of light and dark as our planet pirouettes keeps living clocks closely tethered to the twenty-four-hour day. That connection is crucial for a penguin, parasite, or person to do the right thing at the right time. Life-forms evolving away from the equator also picked up on how the sun's daily arc varied across the year. Changing day lengths signaled changing seasons, which warned of changing hazards and priorities. Circadian rhythms could serve as reliable calendars in addition to daily clocks. For a male humpback whale, decreased day length is believed to prompt its migration south to its winter breeding grounds and perhaps motivate it to begin singing for a mate. For an Arctic fox or mountain hare, the hint of a seasonal shift can trigger a color change. A hare's coat begins transitioning from summer brown to winter white as the days shorten. But now, with the rapid pace of climate change, the snow is melting earlier in the spring. "That mismatch means you can be a completely white rabbit in a brown forest," and your predators can find you, said Micaela

Martinez, an infectious disease ecologist with WE ACT for Environmental Justice.

The same day-length swings will nudge goats and many other animals into breeding season. Some farmers now fool that biology by using artificial lights to raise fertility rates in the natural offseason. It's a strategy reminiscent of the ancient Japanese custom of yogai, in which people tricked caged birds into reproductive maturity—and, therefore, singing—by artificially extending the daily duration of daylight.

I FELL ASLEEP QUICKLY UNDER THE REVOLVING SKY OF STARS, WITH a subtle breeze lapping at my tent's nylon walls. I have often wondered why I can sleep so well camping. Surely my $1,500 fancy foam mattress back home is more comfortable than a thin backpacking pad laid on uneven ground. And my home's insulated walls and thick curtains, which limit distracting sights, sounds, and smells, had to be better than these floppy barriers, right?

Or maybe my body thrilled in removing some of those barriers, even if temporarily. "Humans largely like to think of ourselves, at least since the Industrial Revolution, as disconnected from our natural environment," Martinez told me. The existence of biological rhythms shows that we are enmeshed with the natural world and that this relationship is "very, very important for our health," she said. "The more appreciation we have for that, I think the better we will be as environmental stewards." A greater regard could also improve our health.

For thousands of years, traditional Chinese medicine has emphasized the critical connection between the human body, the natural world, and the cosmos. A timely harmony is the basis for optimal health, according to this practice. In essence, a biological rhythm is a continuous variation in the state of the body in tune with the cycles of the natural world. Vital energy, called qi or chi, circulates through the body and reaches a different organ system every two hours. Energy flow in the heart, for example, peaks between 11:00 a.m. and 1:00 p.m. Traditional Chinese medicine texts caution that detaching from the cycles of day and night or of the seasons can result in disaster and disease.

The ancient Greeks took notice of these natural cycles and their implications, too. They had two words for time. *Chronos* refers to its passage, a *chrono*-logy measurable by an hourglass or a clock's second hand. *Kairos*, on the other hand, is more subjective. It can refer to the opportune moment to do something—plant a seed, eat a meal, or take a pill. Though the Greeks did not explicitly express it, they appear to have understood that a useful clock needed to track elapsed time and be linked with the local sun time to determine the *right time*. "Observe due measure, for right timing is in all things the most important factor," wrote the Greek poet Hesiod in the eighth century BCE. A more famous adage is popularly adapted and credited to the same poet: "There's no place like home."

In the fourth century BCE, another Greek found himself in the right place at the right time—albeit far from home. On his way back from India, Androsthenes of Thasos stopped off on the island of Tylos, now known as Bahrain. There, he noticed that the tamarind tree leaves closed every night and unfurled every sunrise. He made the first written observation of a circadian rhythm. Around the same time, the Greek philosopher Aristotle wrote in his *History of Animals* that bees consistently sleep at night: "Even if a lighted candle be presented they continue sleeping quite as soundly."

Scientists around the world made comparable observations over the following centuries. Yet it took until the eighteenth century for someone to carry out the first documented experiment in the study of biological rhythms. That's when Jean-Jacques d'Ortous de Mairan, a French astronomer, put a *Mimosa pudica* plant in a dark corner of his cupboard and discovered that its fernlike leaves still spread wide and curled inward on a twenty-four-hour schedule. "The sensitive plant thus senses the sun without seeing it in any way," he concluded. De Mairan was a celebrity of sorts. He got at least one nod from Voltaire. He was the first to hypothesize that the sun contributed to the northern lights. Later, he would even have a crater on the moon and a winery named after him. Let's just say that, unlike me, he had better things to do with his time than continue watching a plant slow dance in the dark. A very short journal article in 1729 detailed his findings and proposed future experiments to understand the plant's odd behavior. At the end of the paper, which he had a colleague write for him, de Mairan bowed out and asked botanists and other scientists to take up the case. Many accepted his invitation.

Swedish botanist Carolus Linnaeus made his mark a few years later, noticing that certain flower species reliably opened and closed their flowers at particular times each day. He proposed in the mid-eighteenth century that people could use these floral differences to estimate the time, "even in cloudy weather." Linnaeus went as far as to sketch out how to arrange forty-three flower species into a clock that could approximate times between 3:00 a.m. and 8:00 p.m. The opening of the mouse-ear hawkweed or proliferous pink would mark 9:00 a.m., for example. The closing of the Iceland poppy denoted 7:00 p.m. Goat's beard, also known as Jack-go-to-bed-at-noon, occupied the 3:00 a.m. opening time.

THE 3:00 A.M. ALARM I HAD SET FOR MYSELF IN THE SUNFLOWER field proved unnecessary—unfortunately. Shortly before it would have sounded, horrendous cries of animals shook me awake. A slaughterhouse stood a couple hundred yards away. Disturbed, yet determined not to be derailed, I donned my red-lensed headlamp and crept back into the sunflower maze—this time with more trepidation. I sensed a legit chance I was walking into a horror movie. As I spun my head around, red light danced across the leaves, stems, and heads of budding sunflowers. The bloodred glow added to the eeriness, as did the fact that the plants had moved in the darkness of night. They were leaning in the opposite direction from when I had arrived in the evening. Their heads

now faced eastward toward a faint smear of the Milky Way glowing above the horizon. The sunflowers' leaves were reorienting themselves, too. The front leaves now bowed down so that their tops faced east. The back leaves perked upward for the same effect. Each flower my red beam touched appeared ready for the sun's return. I, however, was far from ready for the new day. I turned around, not having ventured too deep into the maze this time, and hastened back to my tent.

What exactly was happening in that field? If the plants were turning during the night, they couldn't be directly responding to the sun's rays. In other words, the sunflowers couldn't be *following* the sun, right?

The day before I visited Wicks and her sunflowers, I spoke with Stacey Harmer, a plant biologist at the University of California, Davis. I wanted help in understanding the extent of this flower's powers. What was behind its east-to-west daily dance? The question had confounded scientists for a long time. Harmer pointed me to a paper published in 1898 in which the author went to great lengths to determine what compelled the sunflower to move. His experiments, which resembled those of de Mairan and other botanists before him, attempted to eliminate a few plausible external drivers, such as sunlight, temperature, and wind. He went as far as to decapitate sunflowers and found the stems still bowed back and forth across the day. It seemed the plant's dance wasn't all in its head—even if head banging was its signature move.

The overnight reorientation of young sunflowers hinted at the involvement of an innate clock. But most people at the time still presumed that all creatures, those stuck in the ground and those bustling about on or above it, simply set their schedules in response to cues from their surroundings. If the tell wasn't a change in light or temperature, they believed it was probably something else fluctuating in the environment. More than a century later, Harmer and her team followed up on the sunflower mystery. They planted young sunflowers in pots and ran a series of experiments. In one, they let the plants bend from east to west with the sun during the day, but then messed with those same plants at night, rotating some 180 degrees just after sunset. The plants fell for the ploy, continuing to reorient themselves so they faced west instead of east in the morning. In other experiments, the researchers marked the direction potted sunflowers were facing in the field before bringing those pots inside a room with fluorescent light and leaving them there for forty-eight hours. Even with a stationary light source, the plants continued to make their moves.

Harmer and her team concluded in 2016 that an internal clock mechanism must indeed be behind the round-the-clock movement in the sunflower. As I was witnessing in Wicks's maze, the sunflowers were not exactly following the sun but rather staying a step—or bend—ahead of it. The finding added to a lengthening list of organisms we have come to know whose behaviors rely on an internal metronome.

Why would evolution give the sunflower these intrinsic swaying skills? Why not, more simply, program the plant to respond to the sun's light? The researchers' results highlight a few explanations. By getting an early start, the head of a flower heats up faster than if it turned in a passive reaction to the sun's appearance. A warm face attracts more pollinators. There are additional advantages. The direct rays from the sun light up ultraviolet markings on sunflower petals that are visible to bees. Sunrise-facing plants also produce larger and heavier seeds and release pollen for the bees earlier in the morning. Indeed, a clock-driven dance routine appeared beneficial.

The only bees I spotted in Wicks's sunflower fields were bouncing on the heads of oddball early bloomers. A few of these flowers peppered the maze, their stalks towering above the adolescent flowers. Similar spots of yellow popped in another field to the north. These were all so-called volunteer plants, meaning they had germinated from seeds dropped by the previous year's crop and ascended ahead of the more recently planted maze flowers. Wicks was concerned by these more mature plants' behavior. During the evening of my visit, she and I had walked to a cluster of gold in the neighboring field. Stepping carefully over more tall weeds and a few young trees—she also runs a Christmas tree farm on the property—we made our way to the sunflowers in question. Some of the flowers were nearing the end of their days, with petals darkened and drooping. But what bothered her was that they had their backs to the approaching sunset. "They aren't following the sun," she told me. "I don't know what is wrong with them."

Harmer had an answer for that, too. As the sunflower develops, circadian clock genes direct its stem to grow more rapidly on the east side during the day and on the west side at night, swinging its head back and forth. Then, as a sunflower fully matures and blooms, its stem stiffens to support its heavy head, and its youthful nodding slows to a halt. A sunflower stem doesn't harden with the head tipped in just any direction, however; mature flowers fix their heads facing east to the sunrise. Consequently, adult sunflowers always miss the sunset.

AS SOON AS I BEGAN TO PAY ATTENTION, I NOTICED RHYTHMS EVERYwhere. As I stared up from my sleeping bag that night, the only bird above my tent was Aquila, the eagle, a constellation of stars. The night sky at this hour was otherwise quiet, as the living birds slumbered in their communal roosts. I embraced the markedly cooler air now flowing through the valley and through my tent. I also felt an itch coming on, which I pegged to an intermittent buzz I had heard earlier in the night—during the mosquitoes' evening reign. And if my hearing was a little better, I might even have detected the sound of sunflowers stretching taller and broader. Many plants grow fastest late at night, transforming the energy and nutrients they absorb during the day. Henry David Thoreau famously commented on cornstalks inching skyward in the night. Farmers talk about hearing it happen. The audible crackling noise is apparently due to tiny fractures in the plant as it stretches, breaks, repairs itself, and then expands again.

It was a similar story for the scientists at the forefront of chronobiology, the study of time-related phenomena in living things. Once they started looking and listening, they accumulated numerous instances of recurring rhythms. Even Charles Darwin answered de Mairan's call by studying what he called "sleeping plants." Darwin was among a few scientists in the 1800s who came to believe that an internal and inherited mechanism likely underlies plant leaf movements. In the early 1900s, scientists also began to document revolving patterns in animals—fruit flies, rats, monkeys, and humans.

By the middle of the twentieth century, most scientists had grown convinced of a self-sustaining internal timekeeping system in plants and animals that needed only regular winding to stay synchronized with the planet's daily cycles. Much of their rationale centered on findings that an organism's clock runs a bit shorter or longer than twenty-four hours when isolated from external time cues. If an outside geophysical force were solely responsible, those cycles should be precisely twenty-four hours. The idea that an endogenous rhythm should deviate somewhat from twenty-four hours was codified in 1959 by the physician Franz Halberg. He coined the term *circadian*. This internal time varies between species. We now know the clocks in *Mimosa pudica* and in mice naturally run a bit fast, resulting in a sub-twenty-four-hour day, while the average human clock cycles more slowly and stretches the day a little longer than twenty-four hours.

Still, gaps in the science fueled a few doubters. One major sticking point was temperature. While a useful internal clock must be sensitive to certain outside signals to maintain its ties to Earth's twenty-four-hour cycles, the frequency with which that clock ticks must also remain shielded from environmental influences. Temperature introduced a problem. Most biochemical reactions, including metabolic processes in the body, speed up when temperatures rise and slow down when temperatures drop. Man-made clocks were initially inaccurate for a similar reason: increased temperatures lengthened pendulums, which slowed clocks. Could circadian clocks really keep running at a nearly twenty-four-hour clip as temperatures climb and fall across the day and seasons? If a honeybee's clock sped up on a warm summer day, that would spell trouble for Wicks's sunflowers, right?

Colin Pittendrigh, a founding circadian biologist at Stanford and Princeton, eased concerns when he showed that the period of a fruit fly's rhythm was remarkably insensitive to temperature. In his famous experiment, Pittendrigh placed one batch of fruit flies in an abandoned outhouse and another in a pressure cooker submerged in a Rocky Mountain stream. None of the flies could see the daily light-dark cycle. As expected, both groups maintained a daily beat. But despite the great gap in temperature between their environments, Pittendrigh found very little difference in rhythms between the two groups. The temperature compensation of circadian clocks soon became widely accepted. And with Pittendrigh's influence, the "clock" metaphor also soon became widely adopted. To this day, exactly *how* temperature compensation works in living things remains unknown.

By the mid-1960s, Siffre was spelunking without a watch and Aschoff was welcoming volunteers into his timeless Bavarian bunker. More evidence of a curious self-sustaining timepiece kept coming in. Still, to prove its existence, scientists needed to find the smoking gun: Where was this ticking clock, or clocks? A fervent search began.

Given the known power of light to affect the timing of our days, many researchers focused on areas of the animal brain that received information from the eyes. Among the first brains they explored were those of cockroaches and rats—neither of which, I'm grateful to report, crossed paths with me on my reporting adventures. Severing cockroach optic lobes cut the insect's ties to the light-dark cycle and sent their clocks running free and drifting from twenty-four hours. Rats trapped and bred from the alleys of Baltimore experienced a more extreme fate: after researchers destroyed part of their hypothalamus, the rats completely lost the beat.

Scientists homed in closer to the responsible structure in mammals, tracing their way from the retina to the base of the brain. In the front part of the hypothalamus, right above the optic chiasm, is the suprachiasmatic nucleus, or SCN. Damaging only this tiny structure in rats and other rodents resulted in nearly normal animals, except that their once-rhythmic behaviors were now random over the twenty-four-hour day. More scientists built on that research by transplanting a donor SCN into an animal that had no SCN. The recipient took on the rhythm once carried by the donor. The story kept repeating itself. Evidence was strong that, deep inside the brain of these animals, the SCN served as a master timekeeper. And it appeared true for humans, too.

We now know that the SCN in the human brain is formed by a pair of pinhead-sized clusters of roughly 20,000 neurons. Given the average brain contains 86 billion neurons, that's a relatively meager dedication to timeliness. Yet the small bundles pack surprising power and have even charmed scientists. More than one researcher I spoke to referred to the SCN as the "super-charismatic nucleus." The clock generates a near-twenty-four-hour rhythm that coordinates cycles of temperature, digestion, metabolism, and countless other fluctuating functions—including our capacity to be charismatic.

The field of circadian science had come a long way, from discovering the existence of daily rhythms, to determining they were internally driven, to identifying the central timekeeper. Yet more mysteries remained: How did this clock in the brain work? Was it acting alone? While neuroscientists were tracking down the location of the master circadian clock, others were already piecing together the parts that made it tick.

The initial pioneering research on these molecular mechanisms was described in that seminal 1971 paper by Ronald Konopka and Seymour Benzer. The Caltech scientists discovered three genetic typos, each reflecting an abnormal fruit fly rhythm and all mapping to the same gene, *period*. It was the first evidence that genes control clocks. In fact, it was one of the first clues that genes could influence any kind of behavior. Over the following decades, research groups identified several more involved genes and filled in details about how they worked together to create internal rhythms. A series of complex and intertwined feedback loops controlled these biological cogs, gears, and springs—proteins encoded by *per* and other genes break down and build up, inhibit and interact, and compete and coordinate. This push-and-pull process repeats about every twenty-four hours, perhaps in harmony with other ancient temporal processes yet to be untangled.

Fruit flies continued to play a starring role in the research due in part to the simplicity of their genome and short life span, as well as surprising similarities to other animals. But once knowledge of the molecular clock's basic machinery emerged, scientists began recognizing its manifestations and identifying its corresponding components in everything from hamsters to humans. It grew increasingly clear that nearly all life performs this microscopic dance. It's happening throughout your body right now, albeit with a slightly different assortment of genes and proteins than in a fruit fly.

The cells inside of us learned to tell time before we ever consciously could, even before we could cast our eyes on a clock. Aspects of the circadian system develop while we are still in the womb, although obvious manifestations of those rhythms, like sleeping and eating patterns, don't immediately take hold after birth—much to the chagrin of sleep-deprived new parents. Once developed, our body's timekeeping isn't limited to twenty-four-hour cycles. The system also tracks seasonal and lunar patterns, as well as ultradian rhythms that recur repeatedly across the day in periods of hours, minutes, or even seconds. As I wrote this book, scientists at the University of Pittsburgh discovered genes in the brain that follow a twelve-hour rhythm. They also found that many of these rhythms were missing or altered in the brains of patients with schizophrenia, offering a viable clue for risk factors and treatment of the debilitating mental disorder.

THE SIGNIFICANCE OF LIFE'S MOLECULAR CLOCKWORK GAINED MUCH-deserved public recognition in 2017, with the awarding of a Nobel Prize to circadian scientists Michael Young, Jeffrey Hall, and Michael Rosbash. Their combined discoveries, along with those from countless other collaborators, helped reveal this living machinery and paved the way for further breakthroughs in understanding the clocks universal in life across our planet.

Scientists disagree on whether all clocks evolved from one primordial mother or different organisms reinvented the wheel, so to speak. Either way, the "selective pressure to come up with something that could tell time of day goes way, way back," Young told me. Clocks became a key ingredient of life. Scientists have even advocated looking for evidence of circadian rhythms as signs of extraterrestrial life. "If life exists on other planets, it would be reasonable to expect similar evolutionary selection with regard to the rotational period of the planet in question," wrote the authors of a 2005 paper, in which they hinted at a "circadian biosignature" in data from the 1976 Viking lander on Mars. Carrie Partch, a circadian scientist at the University of California, Santa Cruz, concurred: "Surely, life elsewhere adapted to its 'local time.'"

On Earth, some scientists speculate that cyanobacteria were the first to develop this timekeeping tool at least a billion years ago. The single-celled organisms contain chlorophyll, the same photosynthetic pigment that multicellular plants use to harvest energy from the sun. To survive, they need to spend at least part of the day basking in the sun. But cyanobacteria also need to avoid getting fried by the sun's rays. Perhaps some early cyanobacteria evolved to descend deep in the water column to hide from midday's harshest light and to rise closer to the surface around sunrise and sunset. They may have also learned the best times to invest that gathered energy into the production of compounds to defend themselves against lethal ultraviolet radiation.

While cyanobacteria are probably best known, and loathed, today for producing toxic algae blooms, they are widely revered in circadian research labs for yet another reason. The relative simplicity of the ancient organism's clock allows scientists to investigate fundamental mechanisms of circadian rhythms likely found across all kingdoms of life. In 2005, Japanese researchers put three cyanobacterial proteins in a test tube and saw they could team up to track time. Another team of scientists expanded this test tube clock in 2021. The early version could tell time but had no output, or way of communicating information about the time of day to a cyanobacterium. By adding three more proteins and a piece of DNA, the team upgraded it to an entire clock system complete with "hands" to read out the time.

The reconstituted oscillator displayed the three characteristic properties of circadian rhythms: it ticked in the absence of external cues, it recalibrated with information from the environment, and it compensated for outside temperature to maintain a steady beat. The new clock has enabled researchers to study rhythms in real time and ask more questions, such as how the clock regulates the expression of genes and drives downstream manifestations in physiological functions, said Partch, a member of that cyanobacteria research team.

As with the fruit fly and rodent research, the answers they find in cyanobacteria could have implications for us. And for sunflowers. And for all living beings with a clock. It's a long list. "Everything has a clock, even bread mold," Partch said. She has a special appreciation for things growing out of the ground: "I'm obsessed with plants, because they can't get up and move," she told me. "I think that's the clock that we know the least about, even though we started with it back in the 1700s." After all, as I'd learned, it had taken until 2016 for scientists to unriddle one of the most compelling plant curiosities.

AT 6:00 A.M., AFTER ANOTHER SHORT SNOOZE BESIDE THE SUNflower field, I awoke to the dawn's chirping chorus. I zipped myself back out of the tent through a steady drip of condensation. It was now just after sunrise and, under a blanket of fog, I could see that the sunflowers' buds and leaves had barely budged since my last check. They had found their choice positions hours earlier and were still facing me and the sun,

basking in its warm, fog-filtered glow. One morning soon, with yellow petals unfolded, their heated heads would be poised to welcome their favorite pollinators.

The young buds were still nodding east when, later that morning, I attempted the maze. I realized that this natural compass could be advantageous, though less crucial because I could still see over the heads of the flowers. After retracing my steps a few times, I escaped the maze in front of a facade adorned with MA-N-PA'S GENERAL STORE. I then trampled back through tall weeds to pack up camp. I desperately needed to hunt down coffee. Tracking the restless sleep of sunflowers had meant little sleep for me—despite the best efforts of my own circadian clocks. My grumbling stomach hinted that at least some of my inner clocks now thought it was time for breakfast.

As with many scientific discoveries, the revelation that living things have more than one circadian clock, and that those clocks are present in virtually all organs and tissues of the body, came serendipitously—this time, in Ueli Schibler's lab at the University of Geneva, in Switzerland.

Schibler recalled many sleepless nights after one of his doctoral students failed to replicate a former postdoc's remarkable finding in 1990 of a new protein in the rat liver. The student's samples from the animal showed no trace of the protein. Schibler feared his lab might need to retract the resulting high-profile paper. Yet he struggled to believe that his postdoc would have fabricated results. So, first, he decided to take a sample of rat liver himself. This time, the protein miraculously appeared. His concern quickly changed to confusion, and then to curiosity. Finally, an explanation emerged: "It dawned on us that what was really different was the time that these people were working," Schibler told me. The former postdoc who had made the discovery was a night owl. He arrived late to the lab and didn't prepare his samples until after 2:00 p.m. The graduate student, in contrast, grew up as the son of a farmer and was accustomed to waking early to tend to cows. He would get to the lab and take samples from the rats in the morning. Schibler did his sampling after lunch, closer to the postdoc's typical time. His lab followed up on their happy accident by taking liver samples at four-hour intervals and found that levels of the protein varied more than three-hundred-fold across the twenty-four-hour day. Now they had another, perhaps even more remarkable finding to share with the scientific world.

Scientists, including Michael Menaker, one of the early leaders in circadian science, went on to show circadian rhythms in intact tissues and organs—each capable of oscillating on their own. We now know that nearly every one of your approximately ten trillion cells with a nucleus harbors a clock, and that the cellular clocks within each of your organs work together to keep the same time as they direct the rhythmic expression of downstream genes and proteins that do the actual work to maintain your physiology. As the clock genes in your cells turn on and off in regular rhythms across the day, they control the activity of more than ten thousand non-clock genes—roughly half of the genes in your body. You've got rhythm from the hair follicles on your head to the skin on your toes. (Facial hair in men grows fastest during the day, producing the proverbial five-o'clock shadow. There is no 5:00 a.m. shadow if a man shaves at night.) Even your nose abides by a circadian clock. These clocks are connected and tick in special relationships with one another and with the master clock in the brain.

Each of the twenty thousand or so neuron cells in the SCN generates a rhythm, too. This network of neurons is more than the sum of its parts. One researcher showed me a time-lapse video of a slice of SCN taken from a rodent's brain. Luciferase, the enzyme that makes fireflies glow, had been fused to one of the clock genes. This made the neurons light up every time that gene turned on. I watched the cells fade in and out in near-perfect synchrony, brightening and then dimming together over twenty-four hours. The rhythm's amplitude was strong. In another video, individual SCN neurons had been removed from a rodent's brain and placed in a petri dish. These cells lacked connections with one another. A similar time lapse looked like a sky full of twinkling stars.

When I picked out one "star" and focused on it, I could tell that it remained rhythmic as it glowed on and off. But taken as a whole "sky," all rhythmicity was gone. The amplitude flatlined.

Similarly, without the broader network of connections between the SCN and the peripheral clocks distributed throughout the body, the cells of the brain, liver, lungs, muscles, and other organs and systems can lose coordination with one another. Scientists are still untangling just how the body relays clock-syncing messages across neural wires, through blood vessels, and possibly via other means. There are hints that circadian signals from the SCN may even reach the cerebrospinal fluid.

We've heard the SCN compared to a conductor of a peripheral clock orchestra. Yet the SCN may not be the sole conductor, nor the light feed necessarily always the primary cue. (I picture a lightsaber as the conductor's baton.) While scientists initially focused on finding a central pacemaker—looking for a pathway from that critical light input to the brain—it's now clear that other signals can tune other clocks. The liver clock pays close attention to the arrival of food; skeletal muscle clocks take note of when we exercise. A hierarchy may be at play inside us, with the master clock typically at the helm but allowing other clocks to perform solo when biologically beneficial. And it may be a completely different story for plants. As hinted at in that decapitated sunflower experiment, evidence suggests that plants lack a central clock. Instead, an elaborate web of local clocks seems to communicate and coordinate across a plant's tissues and organs.

Much of the focus in the study of circadian rhythms has moved from understanding how our clocks work to unraveling what our clocks control and when they exert that control. This is information we can leverage. We have the power now to estimate what time during the day is best to eat our meals, for example, and what time our brains might be most equipped to tackle problems—like, say, a sunflower maze.

A few weeks after my night with the sunflower buds, I returned with a friend and her daughter to attempt the maze in full bloom. And to reunite with the goats, of course. After visiting Daisy, who was now sprouting small horns, we walked the freshly groomed road to the maze. There, the golden faces of thousands of sunflowers greeted us. We joined the steady flow of families on the paths and added to the stream of vocals: "Another dead end!" "We're lost!" "No, let's go this way!" Most of the flowers had grown from about waist height to my eye level or taller. A happy hum of bees sipping nectar filled the air—occasionally masked by the shrieks of frightened kids. I imagined the bees returning to their hive to shake their bums in a happy dance.

We had arrived in the afternoon. A few weeks earlier, the heads of the young flowers would have been swaying west by this time. Now, nearly every full-grown and heavy-headed sunflower faced east. I caught up with Lora and Ed Wicks, who informed me that one of their neighbors had created a sunflower maze, too, but oriented it in the wrong direction: the backs of sunflowers greeted guests at the entrance of that maze, much like the immature sunflowers had during my first visit here. The only thing the Wickses got backward was the cursive *z*'s in "A-Maze-Zing," spelled out by the winding paths at the end of the maze.

3

Power Hours

Jeff Wilcox has fallen asleep at the dinner table. He has fallen asleep during meetings with his boss and while sitting in the front row at concerts and movies. He was even caught snoring at a sold-out showing of *The Blair Witch Project*. "Don't ask me why everyone else was screaming," he told me. "I thought it was rude to keep me awake."

Staying alert beyond 7:30 p.m. has been a lifelong challenge for Wilcox, a retired construction and real estate entrepreneur living in Oakland, California. "It feels like someone pulls a power cord out of me," he said. Sleeping in past 4:00 a.m. is equally challenging. For decades, this was simply Wilcox's reality. Then, one morning in 2019, he came across a story in *The New York Times* about a rare, but not *that* rare, trait called familial advanced sleep phase, or FASP. The article mentioned a doctor at the University of California, San Francisco, who was studying genetic mutations that affected the pace of a person's circadian clock. That doctor, Louis Ptacek, had found one that sped up the clock so that it ran significantly shorter than twenty-four hours a day, driving someone to fall asleep earlier and wake up earlier than the average person. The description fit. Wilcox cold-called Ptacek and shared his story.

Ptacek asked Wilcox if he saw his unusual sleep timing as a problem. His answer: "Not really." In high school, he made the most of his anomalous morning-ness by fitting in workouts before first period. In college, he chose early classes—almost always easy to come by. And while his friends never understood why he wouldn't go out and party, he stayed out of trouble. When Wilcox later became an entrepreneur, he continued to enjoy the advantages of being an extreme lark. He recalled accomplishing a lot during his uninterrupted mornings. A challenge did arise when he met his future wife, a night owl. Wilcox successfully hid his "problem" from her at the beginning. He took naps before dates. He slapped himself to stay awake on drives home. But he couldn't keep it up forever. Today, he and his wife live on opposite schedules. She'll often go to bed around 2:00 a.m. "If she's not quiet," Wilcox said, "I'll get up and start my day." They joke that they don't really need a king- or even full-size bed; a twin would suffice.

Roughly one in three hundred people have FASP. Many more have other advanced sleep phase disorders. For all these people, the phase of sleep comes unusually early within the daily cycle.

Ptacek was tipped off to the existence of FASP in the late 1990s, when a colleague introduced him to a woman in Utah with a curious sleep schedule. Betsy Thomas would go to bed around 5:00 p.m. and get up around 2:00 a.m. Unlike Wilcox, she found the schedule difficult—even if she reportedly appreciated the quiet of the supermarket during her 4:00 a.m. shopping trips. Other members of her family, including her daughter and granddaughter, shared the same trait. Yet previous doctors didn't take her concerns seriously. Finally, Thomas had the attention of some very engaged experts. She agreed to an experiment far more intense than mine: twenty-plus days alone without access to windows or clocks. From that data, Ptacek and his colleagues deduced that Thomas's internal day measured 23.3 hours long, nearly an hour less between sleeps than the average person. Subsequent DNA tests from family members led the team to pinpoint the first human gene known to alter the speed of circadian rhythms. A mutation in that clock-regulating gene, *per2*, has since been deemed responsible for cases of FASP in multiple families.

This *per2* mutation in humans corresponds to the short-clock mutation in the fly gene, *per*. Another gene, *per3*, and an enzyme that regulates *per*-coded proteins are also implicated in the wide range of clock run times, or periods, from the extreme larks to the extreme owls. Through the course of evolution, humans and other mammals have "diversified the fly's single *per* gene into three," Carrie Partch told me.

Ptacek suggested that FASP long went unrecognized because those affected are "not dropping dead." I'd come to recognize this as a theme with circadian disorders and disruptions, which are usually silent and subtle yet can wield profound impacts on people's lives. Ying-Hui Fu, a human geneticist, soon joined Ptacek in the research. The two have linked FASP with several genetic mutations and anticipate identifying more. Given the large number of genes required to build a biological clock, and the important roles of other non-clock genes in its proper ticking, there is ample room for errors.

The mounting discoveries are helping to solve other mysteries related to clocks, including how rhythms regulate mood and metabolism and how they may be associated with diseases. Still, sleep remains the most widely recognized manifestation of our circadian system's cadence.

SLEEP IS FUNDAMENTAL TO LIFE. MOST LIVING CREATURES SLEEP, ALthough some are less obvious about it than others. Sharks, for one, sleep with their eyes wide open. Some birds sleep while flying. Arctic reindeer may sleep while chewing their cud. And nesting chinstrap penguins sneak in thousands of seconds-long microsleeps across the day, accumulating nearly eleven hours of sleep. Different species also need different amounts of sleep: only about two hours a day for an African elephant or an elephant seal, and up to twenty hours a day for a little brown bat or a koala. An average adult human needs seven to nine hours of sleep.

Still, quantity isn't everything. When we sleep may be just as important as how much.

Sleep duration and timing are controlled by two complexly intertwined yet independent processes in the body: the sleep homeostat and the circadian rhythm. Imagine you pulled an all-nighter to finish a term paper on, say, *Clockwork Orange*. You pushed through the nocturnal hours when the dual processes tried to nudge you to bed. They ramped up your melatonin levels, slowed your heart rate, and cooled your core temperature. They made you sleepy. But you resisted. You may even have enjoyed a second wind around dawn. Finally, at 9:00 a.m., you hand in the paper and beeline to your bed. You're exhausted and very ready to catch up on those lost hours of sleep. Unfortunately, the molecular dynamics going on inside you won't make that so easy.

The sleep homeostat regulates sleep pressure, or how drowsy you feel. It does its soporific magic through a molecule called adenosine, which builds up in your bloodstream while you are awake and breaks down while you sleep. As the accumulating adenosine binds to special receptor sites in the brain, it slows your brain down. Your eyes grow heavy. Theoretically, more than enough adenosine should be around after an all-nighter to fill those sites and initiate a snooze. But your circadian clock at 9:00 a.m. would be sending your biology contradictory instructions. It would be telling you to stay awake, in part by directing your body to produce alerting cortisol and cease production of sleep-inducing melatonin. Even if you did overcome your circadian system's opposition and slip into slumber, you shouldn't expect quality rest. You may not even make it deep enough into REM sleep to experience the disturbing dreams inspired by that book. It's all in the timing.

Ptacek and Ying-Hui emphasized that we can all increase our well-being and reduce our risk, or at least significantly postpone the onset, of dementia and other chronic diseases by improving our sleep. The average American sleeps short of seven hours a night, about an hour less than their grandparents and great-grandparents did in the 1940s. Some of us struggle to fall asleep; others have trouble staying asleep. And not everyone has the luxury of spending ten-plus hours in bed to make up the difference.

The impacts are extensive. Every year in the US, insufficient sleep costs the economy up to $411 billion. These losses are not equally distributed across society. Studies find that Black adults sleep less well and less long than white adults. In two of the poorest cities in the US—Detroit and Birmingham—nearly half of all adults sleep too few hours. Similar to food deserts, where healthy calories are difficult to find, sleep deserts are now recognized as places where good sleep quantity and quality are elusive. The reasons for that are often outside residents' control: noise, light pollution, lack of air-conditioning, irregular work schedules, and feeling unsafe due to higher crime. What's more, research suggests that obstructive sleep apnea, a common sleep disorder that disproportionately affects Black men and women, may meddle with clock genes and lead to circadian disruption.

Scientists recommend we listen to and understand our individual body's rhythms, and tailor our sleep schedule as much as possible to the hours our body is inclined to doze. Seven hours of sleep at the right time can be healthier than eight hours at the wrong time, and not all of us thrive on precisely eight hours. Humans haven't always limited our sleep to one long stretch, either. During the Middle Ages, according to historians, many people split their nights into a "first sleep" and "second sleep"—making use of the in-between period to finish chores, socialize, and procreate. Still, studies suggest that sleep is of a higher quality when it's consolidated into one long segment, with maybe a short supplemental snooze—or siesta—in the midafternoon.

To determine your optimal sleep window and its length, try paying close attention to your body's natural inclinations while on vacation or during other free days. When do you go to bed? When do you wake up without an alarm? With that information in hand, see how much flexibility you can create in your work-life schedule to match those hours seven days a week. In doing this assessment myself, augmented by more than a year of data from various sleep trackers, I realized that I needed to allow myself nearly nine hours in bed at night. The seven and half hours or so I slept during that window of opportunity served me well. I could wake up without an alarm and feel alert. While the particular hours best suited for rest can change with the seasons, my ideal crawl-into-bed and wake-up times in January, when I conducted the audit, were 11:00 p.m. and 8:00 a.m., respectively. As I strengthened my rhythms over the course of the following year, I began to narrow that window and reclaim a few of the wasted, miserable minutes I had spent lying awake.

EACH PERSON'S CIRCADIAN SYSTEM HAS A UNIQUE RHYTHMICITY and alignment with Earth's twenty-four-hour day that depends on both the length of their internal period and how strongly their clocks respond to environmental cues, such as light and dark. The resulting propensity to organize sleep, wake, and other behaviors to earlier or later hours is referred to as a person's chronotype. When the sun rises, are you jumping out of bed? Or are you already hard at work or play? And when the twenty-third hour strikes, is your body prepping for sleep, or is it revving to go for another hour or two—or more?

Chronotypes vary widely among us, much like the length of our feet. Yet society creates even fewer buckets for chronotypes than for shoe sizes. We most commonly hear about just two: morning larks and night owls. Till Roenneberg, a chronobiology consultant and emeritus professor at LMU Munich, added a third broad category—doves—for the approximately three in every four of us whose internal timing falls in between. "They are numerous. But I didn't want to call them pigeons," he told me.

An expanding toolbox allows us to read the hands of our clocks and approximate our chronotype. The gold standard measure of circadian timing is called dim light melatonin onset, or DLMO. A series of samples of blood, urine, or saliva taken under low light conditions during the window around your bedtime can reveal when your body starts to naturally bump up production of melatonin—usually two or three hours before the typical time you go to sleep. That timing is under the direction of the SCN and reflects the circadian phase of sleep.

In the early 2000s, Roenneberg codeveloped the Munich Chronotype Questionnaire as a far simpler, albeit looser, way to gauge a person's chronotype. Roenneberg went beyond the addition of doves to fit people into seven categories, ranging from the extreme early chronotype to the extreme late chronotype. Even seven isn't enough, he admitted.

The full spectrum of chronotypes in the industrialized world plots as a bell-shaped curve with a slight right skew. There are fewer extreme early types, like Wilcox, than extreme late types. Most of us fall somewhere in the middle. Evolution may have favored such diversity; the chance of survival for everyone improves when at least someone is always awake to keep watch for an invading enemy or grizzly bear. Researchers found that members of one of the last tribes to live without artificial light or climate control—the Hadza in Tanzania—were all asleep at the same time for only eighteen minutes a day.

You can estimate your own chronotype by taking the midpoint of your sleep, the core of the Munich questionnaire. As before, think about how you sleep during your days off. During a fantasy three-week getaway to Bali, let's say, you fall asleep around 10:30 p.m. and get up at 6:30 a.m. Your midpoint would be 2:30 a.m., labeling you as a morning lark.

Roenneberg has now calculated the chronotypes of hundreds of thousands of people, plotting them on a bell curve that spans from a sleep midpoint of around midnight for the extreme early bird to 9:30 a.m. at the tail end of the night owls. My midpoint is 3:30 a.m., give or take, depending on the season. That puts me in the middle of the curve and pegs me as a dove. Thank you for not making me a pigeon, Roenneberg.

Although when Roenneberg and I discussed my chronotype, he called me "the most boring intermediate type one can think of." But I digress.

So, what has placed me in the middle of the curve? Is it genes passed down from my parents, my environment, my habits? A predisposition toward the morning or evening persists throughout our lives, explained Roenneberg. This internal inclination will not change "via discipline or will as many people try to suggest," he said. "It's a highly biological trait." If you're an early bird, you might even thank your Neanderthal cousins. One study suggested that when *Homo sapiens* migrated north from Africa into Eurasia, they interbred with the local population of Neanderthals, who were already adapted to the greater seasonal variation in day length at the higher latitude. This interspecies mingling appears to have transferred Neanderthal gene variants that sped up the circadian rhythms of *Homo sapiens*, as well as enhanced the sensitivity of those internal clockworks to resetting stimuli. The clock changes presumably helped the new arrivals synchronize with the northern summer's earlier sunrise. Further backing the biological basis of chronotypes, another study showed that the circadian rhythms of twins most closely matched early and late in life, when decisions about the time to go to bed and wake up are less dominated by outside factors like work or caring for a child.

Yet, as we've hinted, your chronotype isn't all genetic. Your environment and behavior play significant roles. Exposure to the planet's light and dark cycles can push or pull clocks in closer alignment with that earthly day. But our modern world of dark days and bright nights has weakened that contrast and exaggerated our variability. It's artificially shifted and stretched the bell curve. While the days of an extreme lark and an extreme owl are only a few hours apart in the absence of artificial light, they are twelve hours apart in urbanized regions.

The modern chronotype curve has shifted and stretched disproportionately in one direction: toward the owl end. The same person with the same genetics likely leans later today than he or she would have during our ancestors' times. An exception is very early larks, who may be nudged even earlier under modern life's weak day-night distinctions. The midpoint of sleep in preindustrial eras was close to midnight; in industrialized urban areas today, the average is about 4:00 a.m. Midnight is no longer the middle of the night. This could partly explain why I could have a bedtime nearing midnight and a period stretching well beyond twenty-four hours in a bunker, yet still find myself in the middle of the curve.

Other tests helped me further decipher my clockworks. After staying up late spitting into test tubes to assess my DLMO; after plucking hairs from my head and mailing them to Germany for analysis of the phases of gene activity in the follicles; and after sporting a sophisticated watch lent to me by Roenneberg that tracked my activity, sleep, and light exposure over about a month, I began to piece it together. The data and science lessons led me to conclude that while my body rhythms ran longer than twenty-four hours in the bunker, my natural sensitivity to light probably makes me less of an extreme chronotype—otherwise known as "boring"—under contrasting bright days and dark nights aboveground. My Neanderthal cousins may deserve some credit, too.

Without the regular syncing signals from the sun, anyone's internal clocks will gradually drift to some extent. Our behavior can exacerbate the impact. In addition to the clock slowdowns that result from weak illumination across the day, exposure to light during certain windows of time can move the hands of the clock. Some experts refer to a "Netflix effect." Imagine that your circadian period already wants to run a bit longer than twenty-four hours, like it naturally does for most of us when isolated from time-of-day cues. That means you naturally want to stay up a little later every night. And if you stay up late watching Netflix, texting friends, or reading a book, then you expose yourself to artificial light late into the evening. That tricks your SCN into thinking it's earlier than it is and makes it harder to fall asleep. With a late bedtime, there's also a good chance you'll sleep in late and miss the next morning's clock-winding light, which will further confuse your SCN and make it even harder to fall asleep—and even easier to stay up late watching

Netflix—the next night. For some people, however, the Netflix effect can prove a useful tool. Wilcox told me that his clock is easier to hack in the summer than in the winter. He'll frequently take his dogs out for an evening walk while it's still light, which helps delay his bedtime by about an hour—and move it slightly closer to his wife's. Late-in-the-day light may also help elderly people who struggle to stay awake in the evening.

The point is, there is usually some wiggle room in both directions. By waking up earlier and getting more morning light, most people can dial their circadian system forward and fall asleep earlier the next night—the reverse of that Netflix effect. Just a few days of camping without flashlights or other electronics, as studies in the Colorado backcountry showed, can markedly reset clocks and radically narrow differences between owls and larks. During a winter trip, campers began ramping up melatonin production an average of two and a half hours earlier in the night than they had before venturing into the wilderness. The researchers credited the lack of electric light, and the sun. The average camper saw about 14 times more light during the day than they had on a typical winter weekday back home.

As opposed to our weak indoor light environment that artificially amplifies our biological differences, the "very, very strong signal" of naturally bright days and dark nights that our hunter-gatherer ancestors experienced and that we experience when camping "basically reduces those individual differences," said Kenneth Wright, a circadian scientist at the University of Colorado Boulder and author on the camping studies. Unfortunately, pitching a tent every night is not practical for most modern humans.

Complicating matters, body clocks and chronotypes are not fixed. Your chronotype in the summer is likely earlier than your chronotype in the winter. Your chronotype will also change across your life. Young children tend to be relatively early to bed and early to rise. During adolescence, sleep timing can drift later by a couple hours or more. Bedtimes and wake times balance out a bit in adulthood before shifting earlier again in older age—hence, those "early bird specials" sought by seniors.

In our later years, the peaks of rhythms for sleep, temperature, cortisol, and melatonin move earlier but not necessarily at the same pace, putting us at greater risk for desynchrony as we age. The amplitudes of our activity, sleep, and rhythms also flatten. The boundaries between our days and nights blur. Consider the juxtaposition between a young kid who is wired all day and then sleeps hard all night and a grandpa who sits in his chair falling in and out of sleep all day before struggling to fall and stay asleep all night.

Roenneberg emphasized that it's not only chronotype that can change as we age. The relative impact of the sleep homeostat may intensify. "When you're young, the homeostat says, 'I want to go to sleep.' And the clock says, 'Shut up, I'm not gonna go to sleep,'" he told me. "When you get older, the homeostat gets more dominant."

Clocks tick differently between the sexes, too. Higher levels of the female sex hormone estrogen are linked with greater robustness in circadian rhythms. Meanwhile, the male sex hormone testosterone can drive a later chronotype and affect sensitivity to circadian light. Both hormones fluctuate during our lifetimes, which may explain some of the circadian changes we undergo, including the trend for women to display earlier chronotypes than men before middle age.

Whatever our age or sex, many of us spend a good portion of our years living against our internal time. One telltale sign: the use of alarm clocks. Estimates suggest around 80 percent of students and workers wake up with one—probably well before their body clocks would've woken them up. Many people then try to compensate for that lost sleep on days off. Roenneberg and his colleagues coined a term for this familiar discordance: social jet lag. The math for this estimate of how much you live against your biological clock is simply the difference between the midpoint of your sleep on obligation days compared with free days. Research suggests that nearly 70 percent of people in industrialized countries have at least an hour, and a third have more than two hours of this chronic form of jet lag. It's as if we fly multiple time zones west on Friday evening and return east on Monday morning. And then we do it again the next week, and the next, making it extremely hard for our body clocks to get back in sync.

The consequences accumulate. Studies show that the greater a person's social jet lag, the greater their risk of being overweight or obese. Social jet lag is also associated with everything from being a smoker, to heavier caffeine and alcohol consumption, to greater risks of anxiety, depression, cardiovascular disease, and metabolic disorders. It's further linked with poorer cognitive performance and academic achievements. Night owls suffer more social jet lag than morning larks presumably because of the way society is organized, with strict schedules that conflict with their natural rhythms.

THE MOST EXTREME NIGHT OWLS HAVE THEIR OWN CATEGORY. Slightly more common than FASP is delayed sleep phase syndrome, or DSPS. Rather than pulling the timing of sleep earlier in the night, DSPS pushes it later by either lengthening the circadian rhythm or altering the circadian system's sensitivity to light.

Ashley diagnosed herself with DSPS after stumbling across a description online. We connected over Zoom a short while later, just after she had landed her dream job. Ashley told me that she had finally found one that aligned with her body clock. She now works remotely as a veterinary pathologist, reading cases from Australian patients—snakes and such—starting in the early evening, her time. She wraps up in the early morning, goes to bed a few hours later, and rises to start another day by the afternoon. Ashley sticks to this schedule seven days a week, with few exceptions. She's finding that she now rarely needs to set an alarm.

Until the new job, it had been a lifetime of discord. In high school, Ashley's bus would pick her up at 6:30 a.m., a supremely unnatural hour for her. She struggled through college, medical school, residency, and the working world, which typically required her to arrive by 8:00 a.m.

or 9:00 a.m. She would take Benadryl and melatonin to sleep at night, caffeine pills and Monster energy drinks to stay awake during the day. "Even with that, I was always tired," she said. Anxiety and chronic headaches plagued her even when she could get decent hours of sleep.

Ashley's consumption of coffee has now dropped. Her energy has climbed. She says her life is much improved, minus the greater interference with her social life. Finding good times to meet up with friends or go on dates is difficult. She is not alone, as Wilcox's story and others underscore. The Netflix series *Modern Love* aired an episode titled "The Night Girl Finds a Day Boy." It tells the true story of a woman with DSPS who falls in love with a man who lives on dove time. The couple persevere. Happy endings aren't universal for inter-chronotype partners, however. Carrie Partch recalled a former grad student who broke up with a boyfriend because they were "chronobiologically incompatible." The popular dating app Tinder has even added the option to declare yourself an "early bird," "night owl," or "in a spectrum."

Ashley could not point to anyone in her family with the same unusual sleep schedule. She did note that her younger brother works as a land surveyor and, therefore, spends much of his days outside. Could he share the same heritable trait but his daylight exposure masks it? There is scientific precedent. In a study of people who carried a genetic mutation associated with DSPS, Michael Young, the Nobel laureate, identified a few who showed nearly normal rhythms. These were people who saw lots of daylight. Young specifically noted one construction worker with a "very normal sleep pattern." Again, both our genes and the environment can influence the timing of our days.

I asked Ashley and Wilcox whether they'd consider taking a possible future pill that could push or pull their rhythms closer to the social norm. Despite their respective challenges, both were reluctant. "One of the big reasons I got this job, which I'm super happy with, is because they really needed someone to do these hours," Ashley said. Both touted the perk that their hours of greatest productivity typically go uninterrupted. Ashley's working hours now correspond to her peak hours, which span through midnight. Wilcox said he is "running on all cylinders" at six in the morning. However, Wilcox suggested that a pill might still have a time and place: "Maybe I would take it to go to some shows."

WHETHER YOU ARE AN EXTREME LARK, AN EXTREME OWL, OR A COMmon dove, you are not the same person at 6:00 a.m. and 6:00 p.m. As we've learned, nearly every one of the human body's trillions of cells, distributed across organs and tissues, has a circadian clock. These clocks organize our biology and dictate our personal patterns of physiology and psychology—beyond just our sleep cycles.

You can think of your clocks as an automated energy conservation system. The body stores energy from the food we eat, then taps into that energy at optimal times for certain tasks. Countless clock-regulated rhythms regulate a range of energy-intensive processes, from digesting a meal to repairing cellular damage to exerting physical strength. Our bodies can't do it all 24-7. One direct and obvious manifestation of these regular reallocations is our mental and physical energy levels. The average person's alertness reaches a peak a couple of hours after waking and predictably plummets a few hours later—that notorious post-lunch dip. Although a carbohydrate-loaded lunch can compound your sleepy state, the fault is primarily rooted in those two interconnected biological phenomena. Your sleep homeostat and circadian rhythm work in tandem to keep you alert during the day, to prepare you for rest when the day is done, and to ensure you sleep soundly through the night. As sleep-inducing adenosine builds while you're awake, the circadian system directs other physiological rhythms to counter it. But this synergistic programming is not foolproof. The circadian system will fail to fully oppose the rising sleep pressure we feel in the midafternoon. The lull hits us all to some degree, but less harshly with adequate sleep and strongly synced rhythms. The circadian system usually catches up and compensates for the ever-more accumulating adenosine by early evening. Our energy rebounds until hunger for sleep strikes again by bedtime.

As our body's investments in various functions fluctuate during the day, our proficiencies ebb and flow. There are likely best times of day to take a math test, schedule a job interview, perform a piano concerto, conduct surgery, write pages of a book, down a plate of pasta, go for a run, or conceive a baby. Science even suggests there is a best time to soak up sun without getting burned: the morning is generally safer than the evening.

On the flip side of our prime times are our worst times. Most of us will hit our deepest circadian trough between 2:00 a.m. and 5:00 a.m., corresponding to when our core body temperature is lowest. These are the times when alertness plummets and accidents spike. As one scientist put it to me, these are the times when "people are going to be stupid." Name a disaster, and chances are that human error in or around this time window was implicated: the Exxon Valdez accident, the Chernobyl and Three Mile Island nuclear catastrophes, and the Union Carbide chemical plant explosion in Bhopal, India, to name a few. While in the bunker, I watched *Command and Control*, a documentary describing a 1980 incident at another Titan II missile complex a couple dozen miles away in Damascus, Arkansas. Late into a long shift, an Air Force repair team arrived to check vital signs on the missile. While doing so, a socket slipped from a worker's wrench, dropped several stories, ricocheted into the side of the missile, and pierced open its thin skin. Fuel leaked, built up, and culminated in an early morning explosion—thankfully not a detonation of the nuclear warhead itself. Still, one person died and more than twenty others were injured.

Teasing apart the impacts of our sleep homeostat and circadian systems is difficult. Sleep deprivation, regardless of the time of day, is often involved in these tragedies as well. Again, that circadian system–sleep homeostat duality is powerful. Medical errors can spike when the circadian drive for alertness doesn't peak in time to compensate for built-up sleep pressure. For night shift workers, early morning is particularly problematic. For the rest of the population, that post-lunch low between around 2:00 p.m. and 4:00 p.m. can be the most dangerous hours. The AAA Foundation for Traffic Safety warns that a driver going on four to five hours of sleep has about quadruple the risk of crashing compared with a driver who has slept at least seven hours—on par with the crash risk for a legally intoxicated driver. An estimated one in five fatal car crashes involve drowsy driving. More generally, as circadian science would predict, road traffic accidents peak in the midafternoon and the early hours of the morning.

Because our clocks all beat to slightly different drums, these good and bad times naturally differ between individuals, and across the seasons and years. We can learn to read our clocks not only to sleep better but also to live safer and healthier by them. And we need not stop there. Why not take advantage of the emerging circadian science by also timing our days to enhance our productivity and performance?

Take, for example, standardized tests. Studies of schoolchildren have found higher scores on morning tests compared with afternoon tests. But that early edge creeps later as kids enter the night owl years of high school and college. Denni Tommasi, an economist at the University of Bologna, had given circadian rhythms little thought before he spotted a striking trend in an unrelated study involving college students: the average scores on a high-stakes exam were significantly higher at 1:30 p.m. than at both 9:00 a.m. and 4:30 p.m. Tommasi and his colleagues went on to find that the morning disadvantage was far greater for STEM subjects.

Circadian variation across the day, research suggests, accounts for between 20 and 40 percent of the variation in cognitive performance. The implications of these fluxes carry on from school life to work life. Now that I'm more sensitive to my chronotype and paying attention to the timing of my highs and lows, I've discovered that my alertness and optimism crest in the late morning and in the post-dinner hours—on par with the average chronotype. My most productive hours are before noon. Yet I seem to have the most energy for physical activity in the evening. I have since added an app to my phone called RISE, which uses data on my recent activity and sleep to predict these changing tides. The data consistently fit how I feel. I have now made a note to myself to write early and late in the day, and to more or less write off the midafternoon—or allocate those hours for mundane tasks, such as returning emails or emptying the dishwasher.

However, there is an important nuance here. Peaking in alertness is helpful when a cognitive task requires distraction-free focus, such as solving a problem on a STEM exam. It's even been shown to increase the likelihood that we act morally. But certain activities like creating art or making abstract connections can benefit from a little less vigilance and inhibition. In his book *When: The Scientific Secrets of Perfect Timing*, Daniel Pink writes about the "flash of illuminance" that may break through when our thoughts are left unguarded. I've noticed that I lower the barrier to insights and creativity when I tone down my overthinking. It's been my rationale for the occasional glass of wine when I work into the evening. Yet we can also ride our rhythms to these aha moments, which may more readily strike in the afternoon than in the morning—at least for those of us who do our best analytical thinking early in the day.

If the task requires more physical as opposed to mental effort, the equation changes again. To be at your best in the pool or in a pickup basketball game, you should probably aim for the late afternoon or evening, when most of us peak in strength and speed. If you're more interested in the health benefits of exercise, there might be yet another winning time. Morning workouts appear best for reducing the risks of heart disease and stroke, as well as for burning body fat. While we're at it, lifting weights and eating protein in the morning appears optimal for building muscle. Yet exercise in the afternoon more effectively improves blood sugar levels.

The fact that physical aptitude fluctuates across the day is attracting the attention of athletes and coaches who hope to gain an edge over their competition. You might imagine their interest in hearing that, based on one study's calculations, an athlete's performance can vary by as much as 26 percent over the course of the day. For context, during the 2016 Olympics men's one-hundred-meter freestyle swimming event, the fourth place finisher would have won the gold medal if he swam just half of 1 percent—0.24 seconds—faster.

ANDREW MCHILL IS A DIE-HARD SPORTS FAN, ESPECIALLY WHEN IT comes to his favorite professional teams: the Portland Trail Blazers, Seattle Seahawks, and Seattle Mariners. But one thing has always bothered him. Based in the northwest corner of the country, his teams travel more miles during the season than almost any other team in their respective basketball, football, and baseball leagues.

McHill is also a circadian scientist. He loves his work. So, naturally, he'd pondered how travel, whether within or across time zones, might affect a team's rhythms and performance. If a good team had to travel across the country to play a bad team, would jet lag attenuate the difference? "I always thought, is this how the Mariners have been so terrible for so long?" said McHill, a scientist at the Oregon Health & Science University in Portland. After decades of disappointment with the team myself, I appreciated his hypothesis.

For years, he compiled spreadsheet after spreadsheet of data. But isolating the impact of travel from other influences on performance, such as a home crowd taunting an opponent, proved difficult. McHill almost gave up. Then along came COVID-19.

In July 2020, the top twenty-two National Basketball Association teams relocated to the Walt Disney World Resort in Florida, to finish the season in a "bubble"—cut off from the outside world and, ideally, from the spreading virus. When the NBA announced the plan, McHill recalls thinking, "This is the best natural experiment ever." The same teams that had zigzagged the country to play each other prior to the pandemic would now face off without the extraneous factors of home crowds, travel fatigue, or jet lag. And game times would now favor the biological clocks of both teams equally. McHill could finally tease apart travel from talent.

He once again compiled an extensive dataset that included wins, losses, and more detailed measures of offensive and defensive performance across the entire season. Data from games played before the bubble showed a statistically significant edge for the home team over teams traveling across time zones. Precision measures such as how many times a player put the ball through the hoop declined as a team crossed more time zones. On average, players made nearly 2 percent fewer of their shots when traveling across two time zones compared with traveling within their time zone. However, rebounding and other brute force metrics worsened with *any* travel. A good night's sleep appeared the critical factor there.

More curious results emerged as he dug further into the data. Teams traveling west pre-pandemic played worse than those traveling the same distance east. Jet lag wasn't the obvious explanation as the condition is usually worse traveling east than west. (It's easier for most of us to stay up later and wake up later, given our clocks typically run longer than twenty-four hours.) McHill pointed instead to the variable gaps between game time and peak physiology for the players. On average, from late morning through the afternoon, blood flow, blood pressure, and body temperature increase. Muscles loosen and strengthen. Metabolic reactions and the transmission of nerve signals speed up. The body's release of epinephrine and norepinephrine, which aid in acute strength and power, intensifies. Together, these rhythms likely explain why athletic performance normally peaks in the late afternoon or early evening.

Let's say a basketball game tipped off in Portland at 7:00 p.m. If the New York Knicks had traveled west to play the Trail Blazers, then the Knicks would be at a disadvantage. According to their body clocks, the game would start at 10:00 p.m. And it might continue into the early morning hours. If the Knicks were all extreme night owls, then the team might be better tuned for those times. But such a roster would be unlikely. Perhaps this partly explains why the Knicks, as of my writing, have won only fifteen of sixty historic matchups in Portland.

But the balance shifted when the teams moved into the bubble. McHill said he found a leveling of the playing field, or court: "Everyone just played as if they were on their own home court."

The NBA returned the following year to its typical schedule of ping-ponging teams across the country for eighty-two games over six months. At the end of each season, the current NBA playoffs schedule pairs up the top teams to bounce between their respective cities up to five times over each seven-game series. The 2022 finals were a coast-to-coast competition between the Boston Celtics and the Golden State Warriors. The Warriors won the title after six games and three flights across the country. At that point, the Celtics had made the cross-country trek four times.

The impact of sleep loss and circadian rhythm disruption on NBA players during the long season has been called "one of the worst kept secrets in sports." Cheri Mah, a sleep researcher at the Human Performance Center at the University of California, San Francisco, is among health professionals urging the NBA and other sports leagues to address the problem by revising schedules to better support player health and performance. Mah collaborated with ESPN to examine schedules and predict whether NBA teams would win or lose. They didn't base these assessments on the strength of either the team or its opponent, but rather on the intensity of travel, time zones crossed, and recovery time—or lack thereof. The accuracy of their guesses reached as high as 78 percent.

Andre Iguodala of the Warriors was already performing at an all-star level in the mid-2010s when Mah began working with him to remedy his daily schedule. It had been far from optimal: he would snack late into the night (Twizzlers being a favorite), stay up until 4:00 a.m. playing video games, sleep for three or four hours, go to practice, and then nap for two or three hours before starting his routine again. Mah helped him improve both the quantity and quality of his sleep by regulating his bed and mealtimes, as well as his intake of caffeine and alcohol. She even advised him to adjust the temperature of his showers and bedsheets. The changes paid off. When Iguodala got eight or more hours of sleep, he more than doubled his three-point percentage, scored 29 percent more points per minute on the court, and increased his free throw percentage by 9 percent. "I was pretty floored when I saw those stats," Mah told me. Cross-checking his sleep tracker with his game statistics, she said, really helped Iguodala connect the dots for himself, too. That motivated him to comply with his new schedule.

Further evidence of circadian impacts spills from other leagues, including Major League Baseball, the National Hockey League, and the National Football League. One study found MLB teams traveling east—whether coming or going—gave up significantly more home runs than teams traveling west. A thorough review of game statistics put the blame on jet-lagged pitchers. In another study of more than five thousand MLB games with teams playing at different circadian times, the team with the circadian advantage won 52 percent of the time. That winning percentage rose to 61 percent with a three-hour circadian advantage. Similar trends emerged in a review of forty years of NFL games, coauthored by Mah. She and her colleagues looked specifically at evening matchups between East Coast and West Coast teams. Here again, West Coast teams enjoyed a significant edge, beating the Vegas point spread twice as often as East Coast teams. West Coast teams were already known to be at an advantage in Monday Night Football games, a longtime weekly staple of the league. These games typically kick off after 8:00 p.m. eastern time to optimize coast-to-coast television coverage. That may be prime time for the average West Coast player but not so much for the average East Coast player.

The NFL has added more night games in the last decade or two, introducing more of the same associated advantages for the West Coast teams. I now wonder if this might help explain the increased success for my (and McHill's) Seattle Seahawks. The standard Sunday kickoff of around 1:00 p.m., home team time, puts them at a circadian disadvantage: East Coast visitors would be playing at their more optimal early evening time. And any travel would land the Seahawks playing between late morning and early afternoon, their circadian time.

It looks like my local college teams will face more of the same handicap. In late 2023, ten teams from the Pac-12, a major collegiate athletic conference in the western United States, each jumped to join either the Big Ten, the Big 12, or the Atlantic Coast Conference. These latter conferences now include colleges spanning coast to coast. Rather than traveling primarily north and south within a time zone, their sports teams will begin regularly trekking cross-country. More than two dozen scientists responded to the rearrangement, warning of the athletic and academic ramifications inevitable with the coming jet lag and sleep deprivation. One of those scientists, circadian biologist Horacio de la Iglesia of the University of Washington, told me that the playing fields may become more uneven, too—likely tipping in favor of East Coast teams. At the time, the Washington Huskies football team was vying for a national championship in their last season as part of the Pac-12.

Hundreds of sports are played around the world, each relying on a different mix of mental and physical attributes. Certain sports lean more on speed, strength, and stamina; others require greater problem-solving proficiency, hand-eye coordination, and flexibility. And not all attributes follow the same circadian schedule. The performance of a marathoner or other endurance athlete may fluctuate less across the day than the performance of a weightlifter. As far as I can tell, no one has studied the implications for a martial artist, billiards player, or candlepin bowler.

Here's another twist. While sports performance generally peaks in the late afternoon or early evening, that's only an average, representing the middle of the bell curve. When researchers considered chronotype, they found athletic performance maxed out around midday for early birds, 4:00 p.m. for doves, and 8:00 p.m. for night owls. They also found that our skeletal muscle clocks can be pushed forward or pulled backward based on when we exercise. If a dove consistently trains in the morning, for example, they may be able to improve their morning performance.

Even if the goal isn't beating your competition, daily physical activity also helps our body clocks coordinate and maintain robust circadian rhythms. Exercise is just one of many behavioral and environmental signals that can aid in calibrating our circadian clocks. Of course, there is one that rises above them all: the sun.

4

Rhythm and Blues

I had already journeyed to where the sun doesn't shine, dodging daylight in the Arkansas bunker. I have memories and data to support how that screwed me up. But how might my rhythms react to the opposite extreme of never-ending light? To experience the ramifications of darkness deprivation, I decided to embark on a very different and significantly more inviting expedition: a sun-soaked summer trip to Alaska. Up there, around the solstice, the sun only grazes the horizon; night never entirely sets in.

While planning my trip north to the midnight sun, I spoke to Christopher Jung of the University of Alaska Anchorage. He was one of the few scientists in the state researching circadian rhythms, a point that surprised me given the seemingly vast implications at that high latitude. He had just finished a study of how Alaskans' sleep changed across the year. Not surprisingly, participants slept more in the winter than in the spring when the sun rose earlier. This held true even when people used blackout curtains. Jung offered me a few travel tips and then something better: He invited me to join him and a group of local friends—Klint, Sean, Steve, Larry, and Scout, his Australian shepherd—for a late-June weekend of whitewater rafting and camping in Denali National Park.

Without a roof over my head, I would be exposed to daylight all day and all night. Perfect.

A few weeks, an Alaska Airlines flight, and an Alaska Railroad ride later, I was standing in the park. It was midafternoon and I was surrounded by snowcapped mountains that shimmered with the reflecting sun. I went for a long hike, then ate a late dinner. The sun barely moved. It still hovered high in the sky when our group finally adopted a rocky dirt field beside Highway 3 as our campground.

We drank whiskey. We shared stories, mostly my new friends telling terrifying tales of close encounters with bears and moose on past trips—an aspect of this authentic Alaskan adventure I hadn't prepared for. As the hours flew by that night, the sun appeared to steadily float above the horizon. Eventually, the orange ball dipped low enough that it threatened to slink behind the surrounding peaks of the Alaska Range. It was 11:58 p.m. on June 24, and the orb's radiance was waning fast. I had come here for the midnight sun. And I would see the midnight sun, darn it.

It couldn't hurt to get higher, so I climbed to the roof of Klint's RV. He wasn't roughing it as much as the rest of us. And this was no 1970s Winnebago, either, but rather a modern Minnie Winnie. As the digital display of my pink Fitbit turned to midnight, a shallow sliver of sun still glinted atop the mountains, accented a fiery red by hazy clouds above. Jung handed up the whiskey. I took a celebratory swig.

The glow over the horizon never fully faded. After more hours of basking in the dual light of the fire and dusk, followed immediately by dawn, everyone went down for the night. Or attempted to—sleep was elusive. The sun set the orange nylon of my tent aglow. I heard birds chirping at 2:45 a.m. I struggled to turn down the volume in my brain, too. At least the intrusive daylight protected me from stepping in a massive pile of moose poo when I got up to pee in the woods.

The next morning, Jung voiced my exact thoughts: "I could sleep another couple hours. Damn that SCN." Meanwhile, Klint complained about how the towel he had hung over the window in his RV fell off during the night, admitting beams of all-too-early bright morning light. Those of us sleeping in tents did not feel sorry for him.

There was never a dark moment during my five days in Alaska. There was never a dull moment either—from my five moose sightings to a class five rapid. We twice rode the rhythms of the Nenana River in a small fleet of rafts. Its dark blue and intermittently muddied waters carried us leisurely, and then swiftly, and then leisurely again through a series of wide valleys and narrow canyons. On the second ride, I sat on an orange Home Depot folding chair strapped on a blue cataraft. It looked precarious but proved a solid "throne," as we called it. I relished unobstructed views as my experienced guide, Sean, carefully gauged and navigated the waters. He anticipated its moves, setting the raft up for success before every bend and boulder rather than simply responding to the changing conditions—much like a well-tuned circadian clock. Sean was even timely in telling me when I could relax and sip my beer and when I should brace myself with a firm grip on the throne.

AROUND 4.5 BILLION YEARS AGO, THANKS TO MANY COLLISIONS AND collapses during the formation of our solar system, Earth materialized with regular rotations and revolutions. Our planet spun around its axis once every six or so hours in those early days. Eventually, it slowed to the approximately twenty-four-hour eastward rotation we experience today. Meanwhile, Earth continues its annual orbit of the sun. An impact from a Mars-sized planet is widely credited with creating our moon and tilting Earth a bit off its axis, creating seasons and varying day lengths in the Northern and Southern Hemispheres. The moon's gravitational pull has since kept that tilt relatively stable and contributes to the regular ebb and flow of ocean tides that fluctuate across the lunar day and month.

Less than a billion years after Earth's emergence, life began to evolve under these inescapable daily, lunar, seasonal, and annual cycles. Survival meant embracing and exploiting the planet's predictable patterns.

Naturally, organisms developed internal clocks that could generate regular rhythms to match. Entraining to Earth time allowed an organism to foresee and prepare for coming changes rather than react to them. It improved the chances of doing the right thing at the right time. Such synchrony also allowed an organism to revise this "right time" with changing day lengths across the four seasons, especially critical the farther one's home was from the equator.

Again, the clocks bestowed by evolution are not precision timepieces. They almost always run a little faster or slower than twenty-four hours. Like an old grandfather clock, the central clock in our brain and the peripheral clocks throughout our body will naturally drift and, therefore, require periodic calibration to the correct time of day. Thankfully, we don't need a watchmaker for this winding. We just need zeitgebers, German for "time giver." Zeitgebers are rhythmically occurring phenomena that act as cues to keep the body's circadian clocks closely tied to Earth time. Chief among the reliable signals is the rising and setting sun and the corresponding shifts in the color and intensity of light, all by-products of the Earth's spin.

Exposure during the day to the sun's bright, blue-rich light and during the night to only softer, amber hues—resembling those of the setting sun, or a campfire—has helped our clocks tick true for millennia. But it wasn't until the first decade of the 2000s that a series of discoveries ultimately led scientists to look deep into our eyes and find the part of our biology that has long lent light such a powerful sway over our clocks.

As I started looking into this piece of the circadian puzzle, I dug out my mid-1990s high school biology textbook. Sure enough, the standard-issue green and beige *Biology*, fourth edition, by Neil Campbell, discussed only two photoreceptors in our eyes: rods and cones. These cells, which reside in the outer layer of the retina, give us our sense of sight by catching photons of light and converting that energy into electrical signals that our brains then use to build images. Rods, named for their elongated appearance, pick up on subtle changes in light and dark, form and movement. These cells help us see under dim-light conditions. Cones are shaped as you might guess, like cones. With adequate light, they give us our color vision. Different types of cone cells prefer red, green, or blue photons of light. Working together, our cones can distinguish a rainbow of colors—assuming you are neither color-blind nor in a bunker under color-killing red light.

Much has changed since my high school days. I get regular reminders all the time. Still, it was shocking to realize that a fundamental principle I learned in science class is now outdated. A third type of photoreceptor has been discovered lurking in the back of our eyes. These special cells go by a mouthful: intrinsically photosensitive retinal ganglion cells, or ipRGCs. In much the same way as components of the ear allow us to both hear and stay balanced, the eye's rods and cones primarily supply us with vision while ipRGCs perform other useful functions. Among their feats, these cells of the inner retina sense the quality and quantity of light, with additional input from the rods and cones, and quietly transmit that information down the optic nerve into the tiny-but-mighty suprachiasmatic nucleus in the hypothalamus. The master clock reads and translates those notes as it conducts the orchestra of cellular timepieces in the body, ensuring rhythms rise and fall in appropriate relationships with one another and with the external environment. In other words, our visual system paints our three-dimensional physical world; our circadian system adds the fourth dimension, time.

THIS STORY OF DISCOVERY BEGINS WITH A FEW BLIND MICE AND color-changing frogs. In the 1920s, Clyde Keeler, a Harvard graduate student, caught wild mice and bred them in his dorm room as a hobby. He was reportedly part of a community of "mouse fanciers," who traded strains of mice with different coat colors and characteristics "the way that people trade baseball cards," Ignacio Provencio, a biologist at the

University of Virginia, told me. When Keeler later sacrificed one of the mice and sliced open its eye for a school project, he noticed a glaring lack of rods or cones. In fact, many of his mice turned out to be blind. A mutation causing severe retinal degeneration plagued his colony. Yet, despite the mice's blindness, the pupils still shrank when he shone light in their eyes. Keeler had no explanation for the phenomenon other than that an unknown type of light-sensitive cell must be lurking in the retina. (He also noted that a few surviving rods or cones might have been enough to sense the light and trigger those reflexes.) Little did Keeler, the son of a watchmaker, know that he had just happened upon a core element of biology's primary clock winder.

The curious finding sparked more sneers than praise and went mostly ignored for the next sixty years. Eventually, a tag team of researchers revisited Keeler's claim. One of them was Russell Foster, now a neuroscientist at the University of Oxford. He and his colleagues replicated and extended Keeler's study, showing that visually blind mice with retinal degeneration had normal circadian responses to light. But those findings, too, were met with ferocious skepticism. Akin to Galileo's discovery that the Earth orbits the sun rather than the other way around, the idea of a third photoreceptor represented a paradigm shift in science. Foster and his colleagues weren't quite up against the Catholic Church. Still, they did face a generation of scientists not ready to accept a major revision to their doctrine—and potentially a revision to a lot of high school textbooks.

Foster and his team went further in the 1990s, engineering mice to have no rods or cones. These mice still synchronized their rhythms to light-dark cycles. And this time there was no chance that residual rods or cones were responsible. Evidence was building that what meets the eye results in more than just images formed in the brain.

Scientists, including Foster and Brigham and Women's Charles Czeisler, later found cases of blind people without rods and cones who could subconsciously sense light and keep time. One eighty-seven-year-old woman claimed to have zero conscious light detection yet regulated her sleep and wake rhythms just fine. Despite her reluctance—she was adamant that she had no awareness of bright light—she was persuaded by Foster and his colleagues to undergo a test. They exposed her to alternating light and dark, or vice versa, in paired ten-second intervals. When asked to decipher whether a sky blue light was on during the first or the second interval, she went an impressive 33 for 40.

What and where exactly was this pathway that could subconsciously sense light and tell time? So far, scientists had discovered the SCN in a region of the mammalian brain that received signals from the eyes. Scientists had also pinpointed circadian pacemakers in the brains of birds, lizards, and other nonmammalian vertebrates that could directly sense light through the top of the head—a "third eye" of sorts. The pioneering scientist Michael Menaker, who spent most of his career at the University of Virginia, found that sparrows could sync with light-dark cycles, even with their eyes removed. When his research team plucked feathers from the top of blinded birds' heads, the animals rapidly entrained under a daytime light equivalent to less than the light of the moon.

Still, no one had yet identified a photosensitive cell or a light-absorbing molecule that could explain these phenomena. It wasn't for a lack of trying. Ignacio Provencio was a doctoral student with Foster and spent his graduate school years searching for new photoreceptors in the retinas of mice. "I failed miserably," he told me. Later, as a postdoctoral student, Provencio studied a slimier animal: the African clawed frog. Its pigmented skin absorbs daylight, turning darker, lighter, or mottled to match its surroundings. In the late 1990s, Provencio and his colleague Mark Rollag revealed the photopigment behind the frog's camouflage.

The scientists put pigmented cells from tadpole skin, called melanophores, in a culture and watched as the pigment granules inside the cells moved in response to light. In the dark, the granules stayed bunched up around the nucleus of the cell. Because this left most of the cytoplasm clear of the dark granules, the cell looked light colored. In the light, the granules dispersed within the cytoplasm of the cell, giving the cells a dark appearance. Provencio fathomed how this would be adaptive: "A tadpole trying to hide under a rock or a leaf in a stream may have its tail hanging out. The sunlight hitting the tail would cause it to darken, thereby camouflaging it against the dark creek bed." But how did the melanophores detect the light and convert that information into motion of the granules?

The scientists searched the melanophores and eventually isolated the implicated photopigment, with its light-sensitive "opsin" protein. Provencio named it melanopsin (not to be confused with melatonin, the hormone secreted from the pineal gland). He and his colleagues discovered melanopsin in the frog's skin. They also stumbled upon it in its brain and eye.

It was when they saw melanopsin expressed in the ganglion cell layer of the frog's retina, a layer not known to contain rods or cones, that Provencio appreciated that he might have come full circle and found evidence of that elusive third photoreceptor. They looked in the same retinal ganglion cell layer of mice and found the same novel photopigment—confirming Keeler's hunch. Then they saw the same in monkeys, and then in humans.

Other investigators followed the mounting clues and successfully pinpointed the melanopsin-containing photoreceptor cells, the ipRGCs, responsible for the circadian system seeing the light. Finally, they had the eyes and ears of the scientific community. It has since grown evident that the photopigments in rods and cones and the melanopsin within ipRGCs each absorb time-of-day-hinting photons. Their shared excitement in response to a sunlit sky prompts ipRGCs to relay messages to multiple areas in the brain, including the SCN. There, in the master clock, a series of reactions take place. Among the downstream effects is suppression of the pineal gland's release of melatonin, the biochemical signal of night. If you have a healthy, functioning central circadian pathway, your pineal gland will switch back on and pump out melatonin as darkness descends and bedtime approaches.

LIKE MOST OF US, DEREK NAYSMITH TOOK THIS SYSTEM FOR GRANTED during the first three decades of his life. He would get up and go to bed around the same time every day. He would see sunlight, even if filtered through clouds, every day. And he didn't think much of it. Then, one summer evening in 1986, at age thirty-three, he was helping put on a community fireworks display in Edinburgh, when something went terribly wrong. A firework struck Naysmith in the face. His severely damaged eyes had to be removed.

The sleep troubles began almost immediately. At first, he figured it had to do with all the reconstructive surgeries and infections his body endured. As if those ordeals and the loss of sight weren't unpleasant enough, he found himself falling asleep at work and falling more and more out of sync with his wife and two young kids. Over the course of a month, he was waking up later and later in the mornings and getting tired later and later at night. With the loss of his eyes, Naysmith had become time blind. He had lost the ability to regularly reset his inner clocks to the local sun time.

Then, in the mid-1990s, he read a message on a bulletin board posted by Steve Lockley, then a doctoral student at the University of Surrey, recruiting blind people for a sleep study. Naysmith responded. Soon, he had answers for why he felt lousy so much of the time.

After the removal of his eyes, Naysmith was left not only without rods and cones, but without ipRGCs. His SCN no longer had its light feed and began running on its own time. The rhythms his SCN controlled, including his pineal gland's production of melatonin, began cycling nearly every twenty-five hours. He was suffering from what now goes by another mouthful of a name, non-twenty-four-hour sleep-wake rhythm disorder. The only other option at that time to synchronize the SCN in the absence of light was to take synthetic melatonin every twenty-four hours as a replacement time cue. Lockley introduced him to the treatment as part of a Surrey research program and "life became livable again," said Naysmith.

I'm looking out my window right now with extra gratitude. I know that the photons of morning light making their way to the back of my retinas will help me fall asleep tonight. The variety of photoreceptors in my eyes also allow me to see several giant evergreen trees, a blue sky reflecting off Green Lake, and a small bird that has just landed on the roof below. Its red-capped head is turning in sharp, jerky movements as it scouts out its next snack or perch—right on time, I'm sure. My appreciation for color has grown since the bunker, and it was about to get another boost.

A LARGE M&M'S DISPENSER, ROUND AND DECORATED WITH AN EYEball sticker, stands in the hallway outside the Neitz lab in Seattle. Inside the glass jar are by my estimate about thirty thousand red, green, and blue candies, in approximate proportion to the number of respective cone cells in the human eye (not enough blue candies for my taste). Jay and Maureen Neitz are both authorities on color vision who have added circadian science to their repertoire. They use both lenses to trace the concurrent evolution of eyes and living clocks.

Before we go any further, it's worth revisiting that science textbook—a not-yet-outdated section. Photons of light energy travel in microscopic waves. The distance from one wave crest to the next is called a wavelength and corresponds to a color. Short wavelengths lie at the violet-blue end of the spectrum visible to humans, long wavelengths at the orange-red end. Different photopigments in our photoreceptors—whether rods, cones, or ipRGCs—have affinities for different wavelengths of light. You can differentiate a red photon from a blue photon because certain cone photopigments are tuned to absorb more blue photons than red photons and vice versa. Our visual network compares the strength of signals from the stimulated photoreceptors. Almost as if mixing paints on a palette, it will blend this spectral information to create any of millions of colors for our perception. A wavelength of 580 nanometers, for instance, will excite red and green cones nearly equally, telling the brain you're looking at yellow. Certain photoreceptors will also respond to light by wiring clues to the brain about your local time on Earth.

What really excites Jay Neitz goes deeper, to the evolutionary blueprint for our photopigments. The gene sequences for red and green cone photopigments are very similar to each other, but very different from the sequence for blue cone photopigments. A family tree pinpointed when these pigments first appeared as separate molecules: around one billion years ago. Neitz was shocked. Why would such ancient life need to discriminate wavelengths? Back then, the planet was primarily, if not solely, inhabited by single-celled organisms and mostly, if not completely, covered with water. Neitz now speculates that long before the existence of eyes, early lifeforms were using primordial photopigments to register color from their environment, not to *see* the color but to approximate the time. The blue-wavelength light of the sun travels deeper into the water than other wavelengths—creating the ocean's blue hue. These photons signaled organisms to swim down to depths where they were protected from the deadliest beams of solar radiation. When the wavelength balance tipped longer with less-penetrating orange hues at sunrise and sunset, the relative lack of blue light attracted them back to the surface, where they could photosynthesize under less lethal rays.

Neitz believes a daily pattern of swimming up and down in the water column was the first daily rhythm. In this case, organisms were simply reacting to light. But the responsive movement likely buoyed the evolution of an internal timer that allowed aquatic life, including cyanobacteria, to anticipate the sun's position and further optimize their journeys to appropriately avoid or reap its rays—as well as to find food or seek shelter from predators. A rhythmic, color-tuned lifestyle may have laid the foundation upon which evolution constructed clocks in all forms of life. It may explain why sunrise and sunset can be the most favorable times to go fishing, and why many of us feel most energized in the morning and evening.

Color remains a reliable indicator of the time of day. The midday sky delivers a lot of the same blue photons that most efficiently infiltrate ocean depths. These photons include the same short wavelengths, around 480 nanometers, that most excite melanopsin. Further, Neitz has helped map a special circuit in the retina involving cones and ipRGCs that scientists believe is sensitive to the dramatic transitions in color from daylight to twilight and back again. “If I tell you that the sky is orange,” said Neitz, “then you know what time it is to within probably a few minutes.”

Still, Neitz might be in the minority with his emphasis on daylight’s dynamic color changes. While most scientists I spoke with acknowledged the power of melanopsin-stimulating blue, they considered light intensity the more critical time cue for our clocks.

IT WAS A STEREOTYPICALLY CLOUDY PACIFIC NORTHWEST DAY WHEN I visited Russell Van Gelder in his Seattle office, upstairs from Neitz. His desk faced a large window overlooking Lake Union, which took on shades of gray under the sky’s reflection. Van Gelder explained to me that, despite the gray, the range of wavelengths in the sky was comparable to a sunny day. I found this somewhat hard to believe. To my eye, the dull palette out his window closely resembled my parents’ old black-and-white Zenith TV. But Van Gelder assured me that blue photons were still coming through those clouds, just fewer of them.

We can measure quantities of light with a standard unit called lux. Essentially, lux tracks to how brightly we perceive things. Its calculation gives more weight to photons with medium-sized wavelengths, like green, than others. We’ll get into why later. Technically speaking, 1 lux is equivalent to the illumination of a one-square-meter surface one meter away from one candle. A bright sunshiny day can register around 100,000 lux, a cloudy day between 1,000 and 10,000 lux. The full moon maxes out below 1 lux of light, which can be enough for some people to read by once their eyes adapt. Most indoor environments are in the range of 25 to 250 lux.

Van Gelder explained how the visual system evolved to compress differences in light intensity so it could rapidly accommodate extreme contrasts and fluctuations. Early humans may not have walled themselves inside homes and offices all day, but they did wander in and out of caves and shadows. Without the swift shifts enabled by our rapidly firing rods and cones, we would be blinded whenever we tried to look or navigate from inside to outside or vice versa. We would also have little prayer of catching a frisbee, let alone a fastball careening at one hundred miles per hour. Our rods and cones work fast.

Melanopsin, on the other hand, plays the long game. It responds to light far more sluggishly, but once it's turned on, it stays on. Van Gelder likened melanopsin to the Energizer Bunny: "It will keep signaling and signaling and signaling," as long as the light keeps shining. Compared with the highly sensitive photopigments in our rods and cones, the melanopsin in our third photoreceptors also needs light to be thousands of times more intense to be stimulated during the day. "They will respond to that light," said Van Gelder, motioning out his window. Then he gestured inside his office: "This light not so much." Outside, even under cloudy skies, third photoreceptors will register that it is daytime. Inside, standard artificial lights serve as a poor proxy for daylight. They are far less bright, as well as limited in wavelengths. While the visual system responds well to the mid-spectrum wavelengths of light common with indoor lighting, the circadian system much prefers shorter blue wavelengths around that 480-nanometer mark. And that is probably not the only healthy light we're missing. Other wavelengths absent from most modern light bulbs and fixtures, even blocked by our energy-efficient windows, could have substantial effects on our biology, too.

Richard Lang, a basic science researcher at Cincinnati Children's Hospital Medical Center, is among experts uncovering evidence that a violet-sensing photopigment called neuropsin may be important in eye development, wound healing, and metabolism. Unfortunately, violet photons are also hard to come by in our built spaces. Insufficient exposure to wavelengths of violet light around 380 nanometers, which falls at the lower edge of the visible light spectrum, is now a suspect in the rampant rise of nearsightedness, or myopia, an eye disease that usually begins in early childhood. Once a person is myopic, they are also more susceptible to various forms of blindness later in life. The condition was once relatively rare. Today, around two and a half billion people worldwide are myopic, including up to 90 percent of young adults in urban areas of East Asian countries. In contrast, researchers told me they found "virtually no myopia" in Argentinian villages. Studies in children and tree shrews, a species closely related to primates, support the idea that stimulating the eye with violet light during development can suppress the shape changes that result in myopia.

Neuropsin may also be crucial for circadian entrainment of peripheral clocks. Van Gelder and Ethan Buhr, both experts in ophthalmology and circadian science at the University of Washington, found neuropsin in the exposed skin and retinas of rodents and noticed that these organs chose to obey time-of-day signals directly from light via neuropsin over indirect light messages from the SCN. This might serve to more expediently defend the skin against damage from the sun's rays. Scientists had already showed that exposing a mouse to ultraviolet light at different times of day can cause drastically different levels of sunburn and risks of skin cancer.

Meanwhile, yet another curious photopigment is attracting attention. Encephalopsin is activated most efficiently by blue wavelengths of light around 430 nanometers. And its activity appears to be involved in multiple physiological pathways. When Lang and I spoke, he was building a scientific case that a lack of exposure to violet and blue light may be one of the drivers of the silent epidemics of obesity, diabetes, and other metabolic disorders. The research is still very preliminary and leans on animal studies, which don't always translate to humans. Still, Lang speculates that the relative stimulation of neuropsin and encephalopsin helps the body decode the time of day and determine what kind of metabolic activity is needed.

We've only begun to dissect the myriad ways our bodies pick up photons and the places resulting signals travel within us to affect our circadian clocks and to drive other aspects of our physiology. The vibrantly changing wavelengths at daybreak could signal the need to ramp up metabolism for activity, for example, while the color shifts as the sun dips below the horizon could signal a need to wind it down for rest. The dome of the sky is particularly rich in both violet and blue wavelengths at dawn and dusk, noted Lang. Sensitivities to these photons may be yet another tactic that life evolved to harness information from predictable daily changes in the environment. Indoors, we miss all these memos.

AS LIGHTING TECHNOLOGY HAS ADVANCED FROM CAMPFIRES AND candles, to candlelight-colored incandescent bulbs, to cooler fluorescents, and now to light-emitting diodes, or LEDs, we've gotten increasingly good at manipulating our photic environment, and not necessarily for the better. Photons of light are like drugs. A single photon can interact with your body in multiple ways. Its color and intensity matter. But, as with pharmacological drugs, photons can also have varying effects depending on when they strike your body. The timing may be just as, if not more, important than either the wavelengths or the number of photons.

Our circadian system imposes a double standard. If you step into the sun after spending a few hours in a dark movie theater, your vision will rapidly adapt but your circadian system won't suddenly think that you've transitioned from midnight to morning. For this same reason, the light inside Van Gelder's office—or even outside, for that matter—probably did little to alter the phase of my ensemble of clocks during our afternoon visit. But that same light later in the evening could have wildly spun my clocks' hands. The SCN is simply picky about when it welcomes light signals from our ipRGCs. "We are evolved to expect certain types of light at certain times of day," Van Gelder said. "If you don't get that light at that time of day, or you get a different kind of light, it's going to throw off your physiology." This can create vexing situations for us modern humans. Before the advent of artificial lights, however, these clock rules seemed to suit all creatures just fine.

To help me understand light's variable time-of-day effects on the gears of the biological clock, Van Gelder and others pointed me to flying squirrels. More specifically, they steered me to a study by the late Patricia DeCoursey. She was a chronobiologist at the University of Wisconsin–Madison when she published her groundbreaking 1960 paper that described a daily rhythm in the nocturnal animal's sensitivity to light. The flying squirrel, possibly due to its giant light-capturing eyeballs, behaves extraordinarily in rhythm with light and dark cycles. DeCoursey kept her flying squirrels in constant darkness and then administered single ten-minute blasts of light at various times across the animals' subjective days and nights. Light exposures at certain hours caused the squirrels to become more active earlier in the day; at other hours, the blasts pushed their peak activity later. The extent to which the flying squirrel's clock flew forward or backward in time varied with the light's timing as well. DeCoursey plotted her flying squirrel data, creating the first published phase response curve in a mammal, now a standard tool for studying circadian rhythms.

Scientists have since generated many such curves for us diurnal humans. These plots of data have shown that exposure to light in the morning generally advances the circadian clock, meaning you'll get tired earlier in the evening. Light in the evening delays the clock, meaning you'll still feel awake later into the night. Together, exposures at the beginning and end of the day help keep your circadian system stably synced with the sun and calibrated to the changing day lengths of the seasons. Artificial light, however, can deceive. If the master clock receives an unexpected light signal from your ipRGCs after the sun has already set, for example, it will stall in a mistaken attempt to get back into lockstep with the planet's rotation and orbit. If the aberrant light signal arrives before your circadian system expects the sun to rise—a less common scenario for most of us—your clock will jump ahead. Yet the same photons in the middle of the day are expected and, therefore, less likely to budge the hands of your clock.

None of this means you should hide indoors after watching the sunrise or clocking a morning dose of daylight. As is usually the case in science, there are nuances. For instance, getting a lot of circadian-stimulating light throughout the day can strengthen the amplitude and shorten the period of your rhythms, as well as lessen their phase-delaying response to light in the evening. The best predictor of when college students go to bed and when they wake up, according to one study, is the total hours of light they see during the day, regardless of when those hours fall. "The less daylight you get, the later your chronotype becomes," said the University of Washington's Horacio de la Iglesia, an author on that study. The upshot: if you spend the day outside under a blue—or even gray—sky, then you are largely protected from exposure to blue light that night.

As we've seen, our internal pacemakers glean every bit of information they can from our light exposures: the color and intensity, the timing and duration. But light does even more than illuminate our world and tell us the time. Light can also directly affect us through pathways outside of both the visual and circadian systems. Because light is such a good day-night signal, nature has repurposed photoreceptors for "extra duties," said Van Gelder. Evidence suggests that light during the day can immediately increase our alertness, improve our mood, and enhance our smarts. Light's direct effects on our behavior and physiology can also complement the influences from the circadian and sleep homeostat systems, helping us act more quickly and appropriately with changing conditions. But when light strikes or fades at unexpected times, it can promptly override the expression of circadian rhythms—on top of instigating possible clock disruptions. Wildlife and zoo animals showcased this masking phenomenon during the April 2024 solar eclipse: as darkness descended along the path of totality, some nocturnal animals began stirring to life and some diurnal animals began prepping for sleep.

Samer Hattar, a circadian scientist with the US National Institute of Mental Health, shared with me his theory that a separate light pathway works in concert with the circadian system and the sleep homeostat. "There is no doubt that direct light has a huge impact on us that is independent of the SCN," he told me. It seems evolution has endowed us with multiple, sometimes redundant, strategies to pick up on available clues in the natural world. The resulting signals come together to enlighten our clocks and fine-tune our sun-coupled physiology.

AT A LATITUDE OF ALMOST FORTY-EIGHT DEGREES NORTH, US SEATTLEites are closer to the North Pole than to the equator. We see about eight hours of light and sixteen hours of dark in December, and the reverse in June. Those meager eight hours of winter light can also be dim due to cloud cover and the sun's low angle—peaking shy of 20 degrees above the horizon on the winter solstice. And its rays are easily obstructed by the Pacific Northwest's proliferative mountains, hills, and evergreen trees. Friends and family rarely believe me when I tell them I long for the frigid winters in Minnesota, where I went to college. The latitude is not quite as far north as Seattle and the landscape is flatter, opening a vast sky to reflect the region's more regular rays of sunshine. The snow-covered ground augments the effect. One thing I don't miss: squinting my way across campus—a reaction to both the blinding brightness and my eyelashes freezing shut.

Everything I was learning left me wondering how Seattle's huge swings in day length affect circadian clockwork. In fact, Van Gelder and Buhr had run an experiment to answer that very question, at least for mice. Buhr set up a lighting system to emulate a Seattle year consolidated into four months and rigged running wheels to track the animals' activity. During my visit, Van Gelder pulled up the results on his computer. "This was our star," he said, pointing his pen at a figure that showed the activity patterns of one mouse. With twelve hours of daylight, the mouse appeared normal. His activity peaked in the middle of the night, typical for nocturnal creatures. But as the days grew longer and the nights shorter, he continuously compressed his activity in the night until the shortest night on June 21. "Which has to be the worst day for a mouse," said Van Gelder. "They hate long days. They want long nights." At that point, the mouse got off his running wheel for good, went to the corner of his cage, started eating, and got fat. "He basically said, 'Screw it, I'm done,'" said Van Gelder. "Looks a lot like seasonal depression, doesn't it?"

The researchers had messed with this mouse's melanopsin. Variations in the genes for melanopsin are also known risk factors in humans for seasonal affective disorder, or SAD, a form of depression that can recur during the short days of fall and winter. The connection could explain why some people may be more susceptible to the condition and why it might be more prevalent at northern latitudes.

Buhr told me that, as the Seattle days grow short and dark, he will "fight nature" and thoughtfully concoct his own longer day length with the help of various new LED lights that stimulate the circadian system—technologies we'll explore later in this book. The key is moderation and consistency, as augmenting the day with artificial light contributes to circadian troubles. His extensions are primarily bright, earlier mornings. Christopher Jung, the scientist in Alaska, goes to even greater lengths: he had spent much of the winter before our Denali trip working from the less extreme latitudes of Hawaii and Arizona.

THE MID-JUNE DAYS HAD BEEN RAPIDLY LENGTHENING IN SEATTLE as I packed up to head farther north, to a latitude of sixty-three degrees in Alaska. The night before I flew to Fairbanks, where I'd catch the train to Denali, the sun set in Seattle at 9:11 p.m. That same night, where I'd be camping, the sun set at 12:28 a.m. It rose again about three hours later.

I landed midday and went straight from the airport to visit Clay Triplehorn at the Fairbanks Sleep Center. The doctor sees firsthand the consequences of these super-short and very long days on local residents. So, just how bad do Alaskans have it in the winter and summer? The short answer: bad. The long answer: Alaskans' clocks actually struggle year round, Triplehorn told me. In fact, the transitional months might be the worst. He sees the most patients while the daylight hours rapidly lengthen in April and May, and then again as they rapidly shrink in August and September. According to data from the local police department, rates of domestic violence climb during these periods of fast change, too. Rates are also high in the middle of the summer, which Triplehorn speculated is because the extended daylight truncates sleep. Sleep deprivation drives compulsive behaviors and difficulties managing moods. The winter brings its own obvious troubles, including increased rates of anxiety, depression, and suicide. "It impacts things more than we think," he said.

On my last day in Denali, the sun shone above the horizon for around twenty-one hours. As our group again sat around a campfire, the sky suggested it was dusk. Or was it dawn? A thin line still separated the two. We passed around a plate of cheese and moose jerky, homemade by Sean, along with the remnants of the bottle of whiskey, while the guys discussed year-round life in Alaska. Earlier in the weekend, Klint had shared how he chose a flexible job that allowed him to ride his bike while the sun was still up on short winter days. He brought up the unique dilemmas faced by Alaskans again that night. "Imagine going to work at nine, getting off at five, and you go to an office that doesn't have a window. Or you can't see a window," said Klint. "Shitloads of people are like that. Zero sunlight for those people."

The pressure is high to take advantage of precious minutes of daylight in the winter; a similar pressure persists into the summer. "I don't want to miss a moment of this," said Klint, gesturing to our fire circle and the undark sky. "The instant that sun comes out, you start chasing it. You start chasing those sunsets and those little bits of light, and it's later and later and later. And, by the end of June, you're just bullshitting, sitting around at, like, what, one o'clock in the morning?" Our biology had accommodated the extra hours of fun. We were likely experiencing the full spectrum of light's effects: better vision, extra energy, suppressed melatonin, and fooled clocks. The guys frequently used the words *depressed* and *manic* as they described life during the winter and summer, respectively. At one point, Klint likened life in Alaska to being "seasonally bipolar."

THE FIRST SCIENTIFIC CASE REPORT OF BRIGHT LIGHT TREATMENT for recurring winter depression, published in 1982, referred to the patient as "manic-depressive" with a "seasonal mood cycle." The term SAD had yet to be coined. That patient, Herb Kern, had reached out to Alfred Lewy for help while Lewy was at the National Institutes of Health. He and his colleagues were just wrapping up a paper describing how bright light could suppress melatonin.

The received wisdom at the time was that light did not affect circadian rhythms or melatonin levels in humans, as it was known to do in other animals. People still thought our clocks paid most attention to social cues. But a light bulb went off for Lewy, now a psychiatrist at Oregon Health & Science University, after returning from two weeks in Australia and measuring his morning melatonin levels as he adjusted back to East Coast time. Lewy was surprised to find his levels were relatively low the very first morning back, despite his body having just been living on a nearly opposite day-night schedule on the other side of the globe. Could sunlight be suppressing his levels of the hormone? Had scientists been wrong?

The science suggested that the short days of winter drove the seasonal disorder afflicting Kern. So, Lewy thought, what if they could artificially lengthen his short winter days by manipulating melatonin levels with light? The discovery of SAD coincided with the first tests of a successful treatment. Lewy chose thirteen hours as an optimal day length and administered 2,000 lux of light to Kern between 6:00 a.m. and 9:00 a.m., and again between 4:00 p.m. and 7:00 p.m., for ten days. Kern started to respond after the third or fourth day and was more or less feeling good after ten days.

To be fair, the first known use of light therapy to treat the seasonal condition had been administered nearly a century earlier. Captain William Woityra, the icebreaker commander to whom I had sent a sunlamp, tipped me off on this history while on his way to Antarctica in January 2022. He wrote me to say that he had just read Julian Sancton's *Madhouse at the End of the Earth* and recommended that I do the same. The book chronicles an early two-plus-year expedition to the South Pole. An American doctor, Frederick Cook, was on the *Belgica* crew that set sail in 1897. As Woityra pointed out in his email, "they were the first people to winter-over down here." And Cook was the first to describe what he called winter-over syndrome.

As the days rapidly shortened, plunging the crew into months of endless night, the mood on the ship plummeted. No one on board was spared, although some suffered more than others. Sancton describes how even Nansen, the black-and-white cat, went from "preening on deck, purring against the crew members' legs at dinner, or curling up on their chest at night," to becoming hostile and then extremely ill before dying. The men grew sicker, too. Cook blamed the disappearance of the sun and sought a substitute. "Since he couldn't bring the *Belgica* to the light, he attempted to bring light to the Belgica," writes Sancton. Cook ordered men to stand naked in front of the glow of a fire, or what he called the "baking treatment." It seemed to work. The men's moods lifted. Their physical symptoms abated. Of course, the flames of a fire are not nearly as bright as the light used in modern therapies, nor do they contain the blue wavelengths thought to be essential in these treatments. Still, Cook's instincts about the advantages of light were prescient.

Anna Wirz-Justice, a chronobiologist and professor emeritus at the University of Basel, handmade one of the first modern SAD lamps in the 1980s. It consisted of eight stacked fluorescent tubes. "They were so heavy, you needed two people to put them in a car trunk," she told me. Through a series of studies over the following decade, she and her colleagues investigated dose-response relationships for intensity and duration of light exposure. They found that just half an hour of 10,000 lux of light in the morning could send SAD into remission. Further research hinted that timing the light according to a patient's individual circadian rhythm could trigger an even more rapid and optimal response. Years before studying light as therapy, Wirz-Justice's research focused on an apparent paradox: a single night's sleep deprivation could improve mood within hours in patients with major depression. A more recent finding showed that a lack of sleep sensitizes the clock to light, giving that morning pulse more power to synchronize rhythms.

"We have these nondrug methods which we know are amazingly effective," said Wirz-Justice. But neither prescribing an entire night awake nor exposing patients to bright lights can be patented. She lamented that this may explain why there are so few regulations and little research on either promising therapy. Funding is scarce, making it difficult to run a large clinical trial. "However, our effects are so strong that you can actually find them with thirty-five people," said Wirz-Justice. "Whereas, for some antidepressants, they have such a small effect that they need three thousand people to see a difference from placebo."

After forty years of slowly accumulating research, exactly what mix of factors causes SAD—genetics, short days, scant light—and exactly how light therapy works remain mysteries. We know that blue-spectrum light can influence our physiology and behavior via a few different routes, including resetting the master clock in the SCN and its associated seasonal calendar, altering the pineal gland's melatonin production, and directly affecting our alertness, mood, and cognition. But scientists are still hammering out the details, while adding to the list of light therapy's potential benefits, from more robust rhythms to reduced risks for various diseases. Some are also finding that a therapy light need not be overly bright or glary with the right mix of wavelengths. Bottom line: A lack of blues can indeed give you the blues. But supplemental light could provide a lot of relief for a lot of people.

Estimates of the prevalence of SAD range from about 1 percent to 10 percent of the population. Mild SAD may be even more prevalent, with women and people moving from lower to higher latitudes most vulnerable. Genetics undoubtedly plays a role, too, as evidenced by that depressed mouse and further clues: Different rates of SAD were found between populations of Sámi and Finns who migrated to Finland at different points in history. And night owls appear to be at greater risk of SAD and other forms of depression than morning larks.

Whether or not we have a fancy lamp or think we may suffer from SAD, Wirz-Justice recommended that we all seek more light during the day and more darkness at night. "Talk about an easy recipe," she said.

AN ADDED CHALLENGE IS THAT MOST OF US STRUGGLE TO ACCUrately sense how much circadian-activating light enters our eyes. We might think we have a handle on it—just as we might think we can taste the difference between blue and red M&M's. But, again, there is a distinction between how our visual system and the rest of our biology perceive the quantity and quality of light.

Our visual system's response to light is cubic, not linear, allowing it to handle the wide range of intensities in our environment. You sense an eightfold increase in brightness as only twice as bright. Our brains are also not wired to discriminate wavelengths of color produced by artificial lights or screens. What appears a particular color could be a mix of different underlying wavelengths of light manipulated to create that perceptual "color." Even if two lights create the same visual impression, they can have two very different impacts on the circadian system.

What's more, people vary significantly in their responses to light. Researchers found that just 6 lux of light could cut one person's production of sleep-promoting melatonin in half, while another person needed a dose of 350 lux to reach the same level of melatonin suppression. On average, about 25 lux was enough, with light intensities ranging from 10 to 50 lux delaying the rise of melatonin in study participants by 22 to 109 minutes. One scientist I interviewed speculated that lighter eyes may need less blue light to activate the circadian system than darker eyes. Age is another influencing factor. Young children seem to have stronger circadian responses to light than older adults, possibly a reflection of their relatively clearer lenses and larger pupils. Cataracts and the natural yellowing of our lenses as we age can conspire to limit how much blue light gets through to our third photoreceptors. By age forty-five, we have already lost half the circadian photoreception we had in our youth. By age seventy-five, we're down to around 17 percent. And achieving adequate circadian stimulus during the day is even more challenging for people with dementia, as the disease can damage neurons in the region of the brain that controls the clock. Emerging evidence is pointing in the other direction as well—circadian disruption may be a risk factor for dementia.

Knowing now what a poor assessor I am of my light exposures, I decided to enlist the help of LYS, a Copenhagen-based circadian lighting company. I had purchased one of LYS's wearable light sensors before going underground. The quarter-sized black device, made with translucent hour and minute hands to look like a clock itself, estimates the number and type of striking photons. I clipped it on my shirts and sweaters for several months, during and after my bunker experiment. I tried to keep my hair from covering it up, and to remember to relocate it when putting on a coat, with moderate success. A sensor embedded in eyeglasses or contact lenses would provide a more accurate read of the information my ipRGCs collected and sent to my SCN. Nevertheless, the information my LYS sensor sent to the linked smartphone app was enlightening.

When our body needs nourishment, we feel hunger pangs. But we receive no comparable cue when our body craves light. LYS and others aim to fill that gap. The company's smartphone app includes a graph showing hourly light levels throughout the day, with shades between blue and orange to reflect the predominant wavelengths. At the top is a bar that tracks "time spent in natural and healthy light." Two hours is labeled "good." Another bar tracks "time spent in harmful blue light at night." Here, two hours is "high risk." As I quickly realized, days I hardly left the house left me with poor light grades. With a few exceptions. One afternoon, I watched the "daytime" light dial inch closer to "good" as a welcome morning sunbreak streamed a bright beam in through the window by my desk. I resisted the temptation to lower my blinds and temper the glare on my computer screen. Instead, I paused to bask.

The app's graph also labels sunrise and sunset. Before leaving for Alaska, I noted that the sunrise icon was set far ahead of the little alarm icon that represented my typical June wake-up time of 7:30 a.m. The app had made me hyperaware of day length. I noticed that rays of morning sun regularly breached my curtains and woke me up. And I realized that I very rarely witnessed dawn's early light during the summer months. I remedied that in Alaska.

THE TWENTY-FOUR-HOUR DAYLIGHT DIDN'T DETER SCOUT FROM REminding Jung when it was dinnertime. At 6:00 p.m. sharp, the brown-and-white Aussie did circles around us, a dripping tongue dangling from his mouth.

Light may be the primary zeitgeber, but circadian clocks are not solely slaves to the sun. Once the first primordial clocks emerged to anticipate the wild swings in light between day and night, subsequent clock models added supplemental data feeds. The circadian system will take all the information it can get. Just as the first light of day resets the brain clock, the first bite of the day can reset organ clocks. And if Scout, you, or I eat meals at regular times each day, our clocks will direct our body to lower glucose levels and raise concentrations of hunger hormones in anticipation of those meals. Inconsistent meals, especially those that bleed into the night, can weaken rhythms and make a mess of those memos. Scientists now place the timing of food and drink a close second behind light exposure in terms of importance for our circadian system.

Also high on the list is physical activity. Research has shown that exercise can realign and strengthen rhythms in skeletal muscles, and perhaps even throughout the body. In one study, participants who exercised at 7:00 a.m. or between 1:00 p.m. and 4:00 p.m. significantly advanced the phase of their circadian rhythms and the onset and duration of their melatonin secretions, whereas participants who exercised between 7:00 p.m. and 10:00 p.m. delayed their phase and rise in melatonin. Another study showed that exercising at the same time every day bolstered the coordination between skeletal clocks and the SCN, protecting the health of bones and joints.

Temperature is a more complicated zeitgeber. We know that circadian clocks need to compensate for external temperatures to do their job, as the resourceful biologist Colin Pittendrigh had delineated with help from a cold stream and dark outhouse. But that doesn't mean temperature is always ignored. Even small hot-cold fluxes can entrain the clocks of some species—especially in the absence of light-dark cycles. Bees, for instance, can tell time by the temperature cycles inside a hive. Because the Earth's rotation results in environmental cycles of light and dark, as well as hot and cold, the latter might serve as another backup signal. Meanwhile, the SCN drives our internal body temperature to rise during the day and gradually drop in the evening. This internal temperature flux itself can then act as a time-of-day cue for other body clocks.

Extraterrestrial patterns offer yet more clues. Lore that the moon affects human biology and behavior dates to at least the ancient Greeks and Romans. Among the more well-known claims is that a woman's menses follows the moon's cycle. In fact, the term *menses* is tied to the Greek term *mene* for moon. Such lunar powers have been widely dismissed by modern cultures. But emerging evidence of the moon's manipulation of menstrual, sleep-wake, and manic-depressive cycles adds legitimacy to the old myths.

The moon's light intensity varies as its position moves relative to the sun—transitioning from new moon to full moon and back again over about twenty-nine and a half days. The moon's gravitational pull on the Earth also predictably varies as it orbits our planet. A small study published in 2021 found that women's menses intermittently synchronized with the moon's moves. "We hypothesize that in ancient times, human reproductive behavior was synchronous with the Moon but that our modern lifestyle, notably our increasing exposure to artificial light, has changed this relation," the authors stated. A large study published in 2024 adds evidence that life evolved a timekeeping mechanism to anticipate these lunar cycles—perhaps with regular exposure to the moon's gentle tug or illumination keeping this inner circalunar clock calibrated. It's still early days for this line of research, but scientists hope the mounting insights may one day lead to novel treatments for infertility and other conditions.

Other studies link the mood changes of bipolar patients with the gravitational cycles of the sun and moon, and a peak in the incidence of suicide to the week of the full moon. As Shakespeare put it, "It is the very error of the moon; / She comes more nearer earth than she was wont, / And makes men mad."

Meanwhile, Horacio de la Iglesia of the University of Washington has found people living both in urban Seattle and in rural Argentina go to bed later and sleep shorter on nights leading up to the full moon. The waxing moon's increasing illuminance is probably not responsible, he told me, as Seattle's urban glow typically drowns out lunar light. A more likely driver is the gravitational pull or geomagnetic field changes that the moon induces. He and his colleagues speculate that sensitivity to these minute power shifts could be an evolutionary adaptation that allowed our ancestors to take advantage of the bonus hours of visibility gifted by the moon.

The idea that magnetic fields could sway human clocks had been hinted before. Evidence suggests other environmental signals, such as a drop in humidity or a whiff of a pheromone, may have subtle circadian effects for some species, too. Further candidates include social and behavioral factors—from isolation to sex to fear. Even sound makes the list, and not just because loud disturbing noises can upset sleep. One small study found that listening at night to birdsong melodies paired with classical music delayed participants' melatonin production and minimum core body temperature, presumably making them stay awake later.

At the end of the long weekend in Denali, complete with inappropriately timed natural birdsong, I got on the plane back to Seattle. A little after 10:00 p.m., as we made our way south, twilight gave way to night. I joined other passengers gawking out their small plane windows at dark's return. I couldn't help but feel annoyed when bright blue and white lights flashed on inside the plane before we landed in Seattle.

From our great migrations indoors and to higher latitudes, to the odd hours we now eat, work, and play, any combination of today's circadian scramblers can cut us off from the zeitgebers that keep our inner clocks ticking true. "We are essentially walking timepieces," Joseph Bass, a circadian scientist at Northwestern University, told me. "In the wild, animals conform. Their behavior is dictated by an internal timepiece. We are a unique species in that we violate our timing all of the time." Our uniquely human lifestyle has triggered a cascade of consequences.

Part II

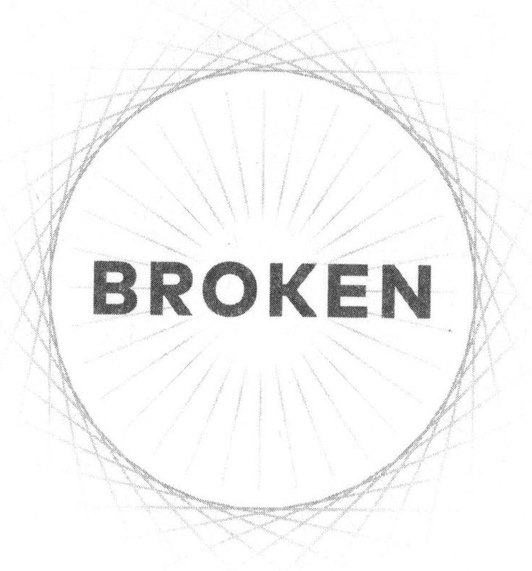

5

Dark Days

"Keep looking up," Andreas Billman urged as we walked the stone sidewalks through the old City of London. Since meeting outside the Barbican Underground Station, we had spied about a dozen "blind windows"—most of which I would have never noticed without Billman as my spotter. Even when looking up, straight at them, they were easy to miss. They blended into buildings. Moreover, each would fail to meet any dictionary definition of a window: none admitted light or air.

Windows have a storied history. The very earliest dwellings didn't have windows, especially those in northern climates. A family might have only punched holes through upper walls or the roof of their home to let in light and let out smoke. Over the ages, people began to construct openings covered with animal hide, cloth, wood, paper, and finally glass. Bricks, not so much—at least not yet. Glass windows eventually evolved into status symbols. Then they became liabilities.

Between 1696 and 1851, England imposed a tax on windows, using them as a crude proxy for property and wealth. When King William III first introduced the tax, his country was in dire need of money to cover the costs of war and new coinage. Prior fundraising strategies had largely failed. A previous hearth tax, for example, required a tax assessor to enter a home and count the number of hearths and stoves. People didn't like that. So-called window peepers could do their job less invasively from the street. Officials administrated the window tax on a sliding scale based on the peepers' counts. The greater the number of windows on a home, it was assumed, the greater the homeowner's wealth. So they levied a heftier tax.

The idea seemed sound, even clever. But the new tax proved imperfect, too. Instead of paying the tax, some building owners blocked up their windows with bricks and mortar. Most notorious for this move were landlords who left poor families in the dark. Then, as architects designed new houses, many began to skimp on those glass-filled openings. They steered clear of bay windows. The once-popular architectural element had become prohibitively expensive. It didn't help that the government had also levied a tax on glass.

Soon, buildings wearing window-shaped divots—bricks inside bricks—filled towns and cities. Many of those buildings remain standing today. Billman and I passed by an eighteenth-century brewery on Chiswell Street that appeared to be among them. I counted twelve blind windows on that brick building. The brewery also wore a plaque noting that King George III and Queen Charlotte had stopped in at least once for a pint during the reign of the window tax.

Samuel Whitbread, the building's original brewer, petitioned for a repeal of the tax in 1829. He had company. The window tax accrued many prominent critics. A character in Henry Fielding's eighteenth-century novel *The History of Tom Jones, a Foundling* complained: "We have stopt up all we could; we have almost blinded the house, I am sure." Charles Dickens added his words of disapproval: "Neither air nor light have been free since the imposition of the window-tax. We are obliged to pay for what nature lavishly supplies to all, at so much per window per year; and the poor who cannot afford the expense are stinted in two of the most urgent necessities of life."

More recently, Wallace Oates and Robert Schwab of the University of Maryland, College Park, dug up and analyzed data from the window tax era, concluding that the tax "distorted" property owners' decisions. The economists looked at records from before and after occasions in which officials augmented the law. One iteration set tax brackets at thresholds of ten, fifteen, and twenty windows. Oates and Schwab found a disproportionate number of buildings had nine, fourteen, or nineteen windows in peepers' subsequent counts. When a revision in 1766 expanded the law to include houses with seven or more windows, the number of houses in England and Wales with exactly seven windows dropped by nearly two thirds. Anecdotal reports back up the data. In a message to Parliament, the president of the Carpenters' Society in London noted that almost every house on Compton Street in Soho had hired him to hide windows.

England was not alone in its unconventional taxation. France, the Netherlands, Scotland, and Ireland all adopted variations of taxes on light. In 1767, after the British Parliament placed a tax on glass—in addition to goods such as tea—windows became an unaffordable luxury for most American colonists. France's tax on windows and doors, which was established in 1798 and remained in place until 1926, was based on both the quantity and size of the windows. The inevitable result: French architects started to design buildings with fewer and smaller windows, especially if a residence was for the poor. In some cases, builders constructed entire houses without windows.

Billman told me he first noticed London's curious windows while riding his bike around the city during the COVID-19 pandemic. He was inspired to capture a selection of them in a photography series he titled *Daylight Robbery* and exhibited at the London Festival of Architecture. While a number of the blind windows he's photographed are likely authentic artifacts of the window tax, others might have been constructed purely for architectural symmetry or other aesthetic reasons. Chimneys or stairwells ran behind a few that I saw. Bricked windows also grew in popularity, and were even adopted in the US to confer a historic look. I understand this fashion trend about as much as acid-washed jeans—even though I once owned a pair.

I later spotted, on my own, many more blind windows around London and Cambridge, and used a rule of thumb from Billman to distinguish the real thing. The newly added bricks didn't always quite match the color or style of a building's original facade. According to Billman, bricks that look "messy" are "more window tax–y."

England eventually lifted the window tax in 1851, yet inequitable daylight robberies continue today. The leading culprit: artificial light.

IN THE YEARS AFTER THE INDUSTRIAL REVOLUTION, AND FOLLOWing the celebrated termination of the window and glass taxes, architects began constructing narrow buildings with larger and more numerous windows and skylights so people could conduct their burgeoning businesses with ample illumination. New York City's Flatiron building still stands tall and thin today, a relic of those daylit days. But designs changed again with the introduction of Thomas Edison's light bulb in the late nineteenth century. Soon, architects relied less on daylight-welcoming windows and more on this new source of tamable illumination. They created deeper buildings filled with electric light. People began to spend more hours of the day inside those buildings. And, after the sun set, they could more easily continue to work and socialize. Day crept further into night.

With this advent of electric lighting a little more than 150 years ago—a mere blink on the evolutionary timescale—we became divorced from Earth's circadian cues. And we didn't look back. Today, the average American or European spends more than 90 percent of the day indoors, often out of eyeshot of a window. That was me for four years in my first office job. When you or I work inside, our daytime light exposure drops by up to a thousandfold and lacks many biologically important wavelengths. We tend to miss sunrise and sunset. Essentially, we live our days in constant twilight and we cripple our biology's capacity to tell the time.

The Old Order Amish are among the rare communities that have retained a natural connection to day and night. Most Amish groups refrain from using electricity from the grid, so the artificial light filling their homes and workplaces is minimal. Digital screens are rare. Scientists studied light exposures during winter and spring for people in an Amish community in Lancaster, Pennsylvania. Compared with the general population, they found the Amish experienced a tenfold greater difference between the intensity of light during their days and nights throughout the year. Similar distinctions have emerged from studies of preindustrial societies in Tanzania, Namibia, and Bolivia. These people live the majority of their hours outdoors.

I don't think I really comprehended this stark contrast in contrasts until Marty Brennan, a daylighting specialist with ZGF Architects, led me on a walking tour in Seattle.

We started with a stroll through Lincoln Park on the shores of Puget Sound. Brennan carried his new spectrometer, a light sensor far bigger and more sophisticated than my LYS button. Every few minutes, he stopped and used the handheld device to assess our surroundings. His eyes lit up each time he showed me the spectral data he collected: 2,811 lux on a wooded trail, 71,584 lux where that path opened onto a rocky beach. Green and red wavelengths of light dominated in the trees; blue wavelengths stood out by the water. "I'm a bit of a spectral snob," he confessed, clutching his work toy.

Along the way, Brennan also referenced what he calls "neural delights," those constantly changing colors and intensities of light, and gradients of temperatures, humidity, breezes, sounds, and smells we experience outdoors. I did notice how much cooler, calmer, and quieter it was standing amid the fluttering leaves in the forest, for example, than when we were gazing out at the dancing whitecaps offshore. "All those sensory changes are powerful to humans," he said. "Juxtapose that with our indoor world, which is highly controlled. Everything is just flat. This whole part of your brain is turned off."

After we left the park, we meandered through neighboring residential streets—by vintage 1920s to 1950s homes—and stepped inside Wildwood Market. There, Brennan pointed his spectrometer down an aisle lined with wine, soda, and chips, which was lit from above by a long fluorescent fixture. The reading: 138 lux of light, mostly in the green-to-red portion of the color spectrum. It was an astonishing difference from what we had seen about a quarter mile away on the trail. Still, despite its low lux, the light inside Wildwood felt bright to me. I hadn't yet learned how our visual system plays tricks on us. The market's lighting was plenty bright to stimulate the visual system, yet its quality and quantity fell well short of what the master circadian clock needs to keep properly ticking and coordinating cellular timepieces in the body. "We are experiencing biological darkness in our indoor environments," Brennan told me. Even though we may be able to see where we're going and can usually avoid tripping and falling, he said, we are inadvertently not stimulating a very old neural circuit in our brains.

Warnings from the medical community about the repercussions of too little daylight date back to at least 1845, during the era of the window tax, with an editorial in *The Lancet*: "Light is as necessary to the perfect growth and nutrition of the human frame as are air and food; and, whenever it is deficient, health fails, and disease appears. This is a fundamental hygienic truth which is apt to be overlooked, despite its very great importance." The editorial also calls out artificial light as a "very bad substitute for natural light."

More recently, researchers have begun to reveal the specifics of our deep, dark predicament. Perhaps the most well-known of the health concerns is a deficiency in sun-made vitamin D, which is estimated to affect more than one billion people globally. But science shows the consequences can extend to circadian, sleep, and other disruptions to our physical and mental health. Too little light during the day can weaken our rhythms and depress our alertness, mood, and health. It can also make us more susceptible to the biological trickery of ubiquitous artificial light at night.

I've read innumerable studies about reduced productivity and performance, and about greater risks of accidents and metabolic disorders. One study of more than eighty-five thousand people in the United

Kingdom linked low light during the day—and, independently, excessive light at night—with major depressive disorder, PTSD, and other mental health problems. Another research team found that animals living with daily shifts between only dim light and darkness had increased beta-amyloid deposition, a hallmark of Alzheimer's disease, compared with animals experiencing stronger day-night contrasts in light. The literature continued to expand as I wrote this book, many of its conclusions echoing those cautions from nearly two centuries ago. *The Lancet* editorial also stated, "A deficiency of light . . . operates powerfully on the mind, depresses the spirits, and diminishes intellectual energy."

The vast disparity between how much light is needed to stimulate the circadian system compared with the visual system has created abundant environments that severely neglect the former—from nursing homes to classrooms. Shadab Rahman, a sleep scientist at Harvard Medical School and Brigham and Women's Hospital in Boston, told me that a typical office setting exposes workers to only about 50 to 100 lux of light—when looking horizontally in a room primarily lit with overhead lighting. Other researchers had a group of office workers wear a device like my LYS sensor to measure their light exposures. It turned out that nearly half of the participants didn't see enough light in the morning hours to sufficiently stimulate their internal clocks, despite almost all of them receiving a recommended level of 300 lux at their desks.

The problem is, most lighting recommendations and energy codes have historically been calculated with only visibility, safety, and energy efficiency in mind. Is it time to revisit how we evaluate light?

IF YOU'RE READING A PRINTED COPY OF THIS BOOK, YOU'RE RELYING on light to make out these words. You could move the book closer to your light source and the page would get brighter—reflecting a higher lux, that standard measure for the illumination of a surface. Lux varies by the distance from the light source, as well as by how much light that source emits. The latter quantity we measure directly at the source with another unit called lumens. Of course, if it's sunlight you're utilizing, then a few inches closer to a massively bright source that is ninety-three million miles away probably won't make much of a difference. But chances are you are indoors and relying on electric light.

At this point, you've probably gleaned that all light isn't equal from the perspective of our biology. The human eye prefers different wavelengths of light for different purposes. Our blue cones like blue wavelengths in a range that peaks around 420 nanometers, and they're okay with the light being somewhat subdued. However, because a cone's action is fast and brief, the blue cone input to the circadian system is strongest when there is a contrast, such as when accompanied by long-wavelength light at sunrise and sunset. Then, as our eyes move around to take in the dynamic view, the circuit can refresh and send repeated signals to our brain. Meanwhile, as we've learned, the melanopsin in our ipRGCs will trigger a potent circadian signal only in response to very bright and prolonged blue light, ideally concentrated around 480 nanometers. Photons of other wavelengths can elicit the same response; it just takes a lot more of them. So, perhaps via multiple pathways, blue light will excite and prompt our third photoreceptors to convey information to our circadian system about the approximate local Earth time.

When it comes to reading a book and, generally, seeing the world around us, our visual system favors wavelengths in the vicinity of 555 nanometers. As you might remember from Neitz's M&M dispenser, we have far more green and red cones than blue cones. And our green and red cones are most excited by wavelengths around 530 and 560 nanometers, respectively. Together, their collective sensitivity peaks at the yellowish-green 555-nanometer mark. This is why an international committee determined a century ago that, for measures of indoor lighting such as lux and lumens, greater credit should be given for wavelengths of light that are closer to 555 nanometers.

Lighting designers today still use this green-weighted calculus when determining the number of lumens needed to meet a certain lux standard in a building. The critical point here: short-wavelength blue light and long-wavelength red light have largely become, by the definition of lumens, wasted energy. As a result, essential parts of the spectrum end up omitted from the indoor spaces in which we spend most of our lives.

Naomi Miller, a lighting scientist at the Pacific Northwest National Laboratory in Portland, Oregon, was among multiple experts I spoke with who argued passionately that energy efficiency standards need to consider that lighting does more than just enable us to see. Most specifically, she and others said, we need new lighting definitions that account for the wavelengths preferred by our circadian system and which have been largely ignored and widely absent from our indoor environments. "Your body has evolved to use pretty much all of the daylight spectrum for one purpose or another," she told me. "It doesn't make sense that we would only make use of the wavelengths that count towards lumens."

In 2022, a team of circadian scientists scoured the literature and made recommendations based on a new measure of lighting that accounts for these important nonvisual effects: melanopic equivalent daylight illuminance (EDI). Melanopic EDI is one of several alternative measures devised over recent years to estimate melanopsin-stimulating light. It is both a measure of intensity and of color. While lux, lumens, and other traditional measures reward light focused around the green part of the spectrum, melanopic EDI values go up with greater light concentrated around blue.

The scientists encourage the use of daylighting and emerging tunable LED technologies to achieve the proposed benchmarks inside buildings, including their recommended minimum throughout the daytime: a melanopic EDI of 250 lux, measured at eye level. Meeting these metrics would reintroduce a more marked day-night signal for occupants, ultimately promoting stronger circadian rhythms, improving daytime alertness, and supporting sleep. "Electric light is one of the most amazing inventions that humans have made. It has transformed our lives," said University of Colorado Boulder's Kenneth Wright, one of the coauthors on the recommendations. "But we've realized that it has also led to some negatives."

We can expect the benchmarks to evolve with the science. Each lighting target was also determined only for an average healthy adult. Few of us are precisely average. Older patients in a nursing home may be less sensitive to light and may need a higher melanopic EDI to achieve the same effect. Even people of the same age can differ significantly in how they respond to the same light, possibly based on sex, eye color, or other not-yet-identified factors. We also still don't know how long we need to bask in 250 lux during the day, nor how much the precise timing matters. We don't even know whether this benchmark is high enough for the average person. We simply lack enough real-world data to be sure of anything.

IN ADDITION TO THE UBIQUITOUS BRICKED-UP WINDOWS, MY WALKS in England led me to another relic of early battles over light. I spotted my first ANCIENT LIGHTS sign in the alleyway beside the St. Christopher's Inn pub near London Bridge. It hung on a brick wall, beside a bronze lantern, and between strings of Christmas lights. When I asked the bartender inside about the sign, it was the first she'd heard of it. She had never noticed its presence. I saw a second sign just a few blocks from my Airbnb by Buckingham Palace, on what is now the Burdett-Coutts & Townshend Foundation CE Primary School, and yet another on a stroll through Cambridge, posted on the aged bay window frame of a home on Pemberton Terrace. On that same street, I spied a brick building with five of seven windows on the front side blocked up.

So, what were these ANCIENT LIGHTS signs about? Many of the placards looked relatively new, and none of them were displayed next to any obviously historic lamps. That shiny bronze lantern by the pub was clearly modern vintage. However, the signs were all adjacent to at least one window—the kind made of glass, not bricks. As I learned, they were vestiges of an English property law dating back to 1663. Its current form hails from 1832. The law gives homeowners who had enjoyed natural light via a window continuously for twenty years the right to continue receiving that "ancient light." It does so by prohibiting those with adjoining land from obstructing it, whether by raising a wall, erecting a building, or even planting a tree in front of the safeguarded windows. Construction of the Chelsea Football Club's new Stamford Bridge stadium was almost aborted in 2018 after a London family attempted to leverage the nearly five-hundred-year-old law. The Crosthwaites had lived in their home, a reported "dropkick" from the soccer stadium, for fifty years.

I discovered that the concept of preserving daylight dates even further back. The Roman emperor Justinian declared that no neighbor could block light that had been enjoyed for heat, light, or operation of a sundial. And the ancient Greeks recognized and protected the human right to light, orienting streets and buildings to take advantage of the sun's rays. With the same goal in mind, regulations in eighteenth-century Denmark limited building heights to the width of streets.

The right to light remains upheld in some parts of the world. A building in Bali, Indonesia, cannot rise above fifteen meters, or about the height of a coconut tree. The rule helped to thwart the Trump Organization's attempts to construct a resort tower on the island. In China, some residential buildings in cities with populations of a half million or more must receive at least two hours of sunshine on the dahan day, which falls around January 21. In Japan, nisshōken, literally the "right to sunshine," governs the height and shape of buildings so they don't block sunlight from reaching other buildings for more than a set number of hours a day. Paris reinforced its historic standing as a low-rise city by adopting a twelve-story limit for new buildings. And in Zurich, Switzerland, regulations restrict the time a high-rise addition in a residential area can cast shadows on a surface to less than two hours a day during the winter. This, by the way, is in the same country where people with depression are prescribed morning walks.

The urban fabric of many old European cities invites the open air and blue sky. I relished the wide streets and squat buildings where I stayed during my weeks in London, Copenhagen, and Amsterdam. More daylight streamed through—onto sidewalks and into offices, cafés, and homes (and Airbnbs)—than I was used to in Seattle. Even on overcast days, it felt relatively bright and warm. Copenhagen's vibrant foliage-shedding trees and pastel-colored buildings didn't hurt the vibe.

In many places, right-to-light aspirations have faded over time. Even longtime holdouts like London and São Paulo have slowly begun to cave in to pressures to raise their height limits. These are tough decisions between competing evils. A lack of affordable housing is a major issue in many cities, and taller buildings generally mean lower rents. But the increased density brings decreased daylight, which is regularly discounted in these deliberations. As one Polish architect put it to me, "Daylight is free. If something is free, people kind of neglect the importance of it."

As more people around the world crowd into cities, and as more homes and offices are squeezed into smaller spaces, our collective access to daylight both indoors and outdoors further dwindles.

THE LEGAL PRINCIPLE OF ANCIENT LIGHTS NEVER QUITE TOOK HOLD in the US. The country adopted it at the time of independence, but a New York court decision in 1838 began the doctrine's downfall. American economic expansion and notions of private property would take precedence: "It may do well enough in England," the court decided. "But it cannot be applied in the growing cities and villages of this country, without working the most mischievous consequences." By the first part of the twentieth century, high courts in all states had rejected ancient lights, citing this case. Michael Kwartler, an architect and urban planner in New York City, distilled for me the cavalier attitude extolled by the US courts: "This is America. We have all this space and land so, you know, what's with this right to light? Nobody's ever going to build that close together, right? Go figure."

Momentum revived in the twentieth century, as recognition rose for the consequences of building high. A 1913 headline in *The New York Times* read MANY CITIES LIMIT BUILDING HEIGHTS: RESTRICTIONS IN EUROPE SHOW THAT SKYSCRAPERS ARE NOT POPULAR THERE. GREATER LEEWAY HERE. The article goes on to describe how a few US cities, including Boston, Baltimore, Denver, and Los Angeles, had begun insisting that "their buildings and streets conform to the highest standards not only of usefulness but of beauty, safety, and healthfulness." New York City eventually followed with a 1916 zoning law that didn't exactly restrict heights but did mandate that tall buildings include setbacks from the street to allow access to the sky. This, I learned, explains why the Empire State Building resembles a giant wedding cake. Its early 1930s construction followed controversy over the 1915 arrival of the Equitable Building in Lower Manhattan. That colossal building lacked setbacks, rose more than five hundred feet, and cast a seven-acre shadow over the neighborhood.

Still, the problem did not go away. Not in the least. A 2016 study by New York University researchers found that shadows cover most Manhattan neighborhoods for at least half the day. The darkness spills onto the city streets and sidewalks, and into parks, discouraging hours once spent on sunny stoops, gardens, and lawns. The Equitable Building is now just one of many behemoths casting shadows on the intersection of Cedar and William in Lower Manhattan, deemed one of the darkest in New York City. It didn't disappoint when I strolled to that corner during a visit in December 2021. Despite a clear blue sky and a sun near its zenith, shortly before noon, few photons reached street level. My sunglasses remained perched on the top of my head.

Where there are dark streets and sidewalks, there are also dark living rooms and bedrooms. Kwartler introduced me to a befitting British term: the "grumble line." Assessing the right to light in England involves establishing a threshold in a room beyond which light is inadequate to read. In other words, this is the point in the apartment where "you begin to grumble," he said. One of Kwartler's early homes in New York was a ground-floor apartment in a row house on the Upper West Side. "I never knew what time of day it was until I had turned on the television or the radio. I mean, it was always the same gloomy," he told me. "I would sleep late or get up too early. It was so depressing."

It's a common scenario for New Yorkers and a reason many flock to their shared backyard: Central Park. I spoke with photographer Andrew Brucker, who has been one of these urban sun seekers since his move to the city in 1978. These quests are a bit more difficult than they once were, he told me. In early November 2018, Brucker was walking through Sheep Meadow, a large expanse near the southern boundary of the park, when he noticed a "gigantic ominous shadow" across the lawn. He snapped a photo with his phone. It shows the silhouette of a newly rising Central Park Tower cast into its namesake park, with parkgoers dodging the dark and lying on the grass to the left and right of the shadow. The slim skyscraper was still under construction at the time and hadn't yet reached its full height. "I don't tend to hang out on that end of the park anymore," said Brucker, who is known for his portraits of high-profile actors, musicians, and writers—some of whom may now reside in this or neighboring high-profile towers.

AS OF THIS WRITING, CENTRAL PARK TOWER IS THE TALLEST RESIdential building in the world. It ascends 1,550 feet above New York City. Its homes boast floor-to-ceiling windows and, on the upper floors, zero obstructions to daylight—minus LaGuardia-bound aircraft and migrating birds. On the website for the now completed Central Park Tower, I found a photo of a blond woman in a black robe basking in a sunbeam on her bed. Light reflects off a very large diamond ring. For upward of $20 million, you too could occupy one of the residences that span some of the building's high-altitude floors.

Several new towers were popping up across Manhattan when I visited in late 2021, particularly along the southern end of its most famous park—a stretch dubbed "Billionaires' Row." At their top, spectacular penthouses were filled with unfettered natural light. Below, more-affordable homes and public parks fell in their shadows during the day and bore the brunt of artificial light at night. It was a night-and-day contrast between people living up high versus down low—literally and financially.

The shadows of daytime darkness have long fallen heaviest upon the urban poor. In his 1890 book, *How the Other Half Lives*, photojournalist Jacob Riis used flash photography, a new technology at the time, to shed light on the plight of the poor in New York City. A host of horrors plagued residents of crowded tenements: malnourishment, disease, extreme temperatures—and a lack of daylight. Riis described the narrow streets and alleyways between towering tenements. He wrote of feeling his way through dark stairwells, hallways, and bedrooms. While society at the time recognized sunlight as important for health—a tenet even recited by schoolchildren—the precious commodity was in scant supply in these neighborhoods. Through his work, Riis raised awareness of a need for greater and more equitable access to light and air. Minor improvements followed. Yet again, the burden was never fully lifted.

Around the world today, people are voicing concerns about diminished access to daylight. Railroad-style apartments, where air and light enter at only one end, were traditionally found in older tenement buildings and still house people in large cities today. They are among the more affordable options, aside from basement apartments and strategies like splitting a two-bedroom apartment between four people, as I did as a grad student in Manhattan.

New York City residential building codes call for ten square feet of window for every one hundred square feet of habitable living space.

And every bedroom must have a window. Of course, people regularly sidestep the rule, converting windowless living or dining rooms into bedrooms as they squeeze four people into a two-bedroom apartment. In my hometown of Seattle, codes call for artificial light with an average illumination of 107 lux or eight square feet of window for every one hundred square feet of living space. Marty Brennan called that "pitifully low." He suggested the window area should be equivalent to at least 12 percent of the living area.

Construction projects for public housing frequently stop at these bare minimum standards, while many marketed homes will respond to house hunters' demands for daylight. Brennan noted how new townhomes in Seattle are generally narrow, with three sides of windows, and usually exceed code minimums so they can sell for maximum prices. Suburban homes are on yet another level, with four sides available for windows and ample square footage for skylights.

Workplaces and schools follow a similar pattern of inequities. Some countries around the world, including Germany, promote or even require direct access to daylight for office workers. That is generally not the case in the US, where it is often only executives who enjoy a windowed office. Even when a daylight requirement might exist, there may be exceptions for factories and warehouses. It's a similar story for schools.

Windowless classrooms became increasingly common in the 1960s after questionable studies implied that windows distracted students, notes Lisa Heschong in her book *Visual Delight in Architecture*. The rationale at the time also included energy savings and civil defense. Sure, a lack of windows cut the risk of "flying glass and blinding light" from a nuclear blast, writes Heschong, a world-renowned daylight expert based in Santa Cruz, California. But the repercussions of blocking daylight and views have since plagued generations of kids. In 1999, Heschong authored a study finding that elementary school kids exposed to ample natural daylight learned faster and performed better on standardized tests than students learning solely under electric light. Many studies have since added to the concerns, including poorer sleep for students at schools lacking sufficient short-wavelength light.

If you're lucky enough to have ample unobstructed windows in your office, school, or home, your real estate of glass still does not guarantee adequate circadian-stimulating light. Somehow, we keep finding ways to make it harder on our clocks. Privacy issues, heat from the sun, or glare on our computer screens can compel us to lower our blinds. And there's another, more invisible barrier that few people know about. At least it was news to me.

HESCHONG REMODELED AN OLDER HOME IN SANTA CRUZ THAT BRIMS with windows—some old, some new. One easy tell of the difference: her plants "clearly suffer" in front of the newer panes of glass, she said, and thrive in front of the older panes. Modern energy-efficient windows keep out the sun's heat by reflecting away both ultraviolet and infrared light. Yet most plants need broad-spectrum light, explained Heschong. And tiny insects tending to those plants need broad-spectrum light. "Work your way up the ecological chain and, logically, there's a good chance that may be true for humans, too," she told me.

Since their introduction in the 1980s, new generations of low-e (for "low emissivity") window coatings have progressively narrowed the band of transmitted light ever closer to only the visible portion of the spectrum, focused around that 555-nanometer mark favored by daytime color vision. The idea is to maximize visual performance and minimize energy loss through the glass. Building codes now typically mandate these highly energy-efficient windows. It's all well meaning, of course. Saving energy as we confront the concerns of climate change is crucial. Just like with the light emanating from our electric bulbs, it made good sense to concentrate the light coming in through our windows to those visible wavelengths. But, given what we are learning now, we may want to reconsider.

Lost light from both ends of the spectrum may undermine public health, noted Naomi Miller. In addition to the powerful sway of short-wavelength violet and blue photons on our physiology, researchers are uncovering apparent beneficial impacts of long-wavelength red light. Interest has grown since the 1990s, when NASA scientists noticed wounds on their hands healed faster while working with plants grown under red LEDs. We now know the same photons can stimulate hair growth. And animal studies hint that near-infrared light, which lies just past red and beyond the visible spectrum, could protect against retinal disease, Alzheimer's disease, and other aging-related problems. There is much left to sort out, but Miller expects scientists to continue uncovering benefits of these wavelengths. As she reminded me, humans evolved with the full spectrum of sunlight yet we now spend our days under only slices of that spectrum: "Most window glass sold for the last thirty years is not providing the exposure we need."

Thomas Culp, a glazing expert and energy code consultant in La Crosse, Wisconsin, taught me a simple way to determine the type of glass used in a window: Shine a smartphone flashlight into the pane (or panes) of glass. A colored reflection—even a pale blue, green, or red—on the front or back of each plane suggests a low-e coating on that surface. When I tried the flashlight trick on the window by my computer desk at home, I saw one round green dot. I also found specification etchings on the glass. I snapped a picture and sent it to Culp. "You got lucky," he wrote back. I had low-e glass on the outside pane, with a "double silver" coating on the second surface, clear glass on the interior, and an argon gas fill. He called this "good." In other words, I still had a decent amount of blue light penetrating my apartment office. The energy-efficiency coatings likely filtered out only half to two thirds of the blue and violet photons. This sounded sizeable. But, as his data show, the newest triple silver low-e coatings on the market today can intercept 70 to 85 percent of the circadian-stimulating light.

Clearly, the easiest way around these roadblocks is to spend less time indoors. We can eat breakfast or lunch outside. We can walk or bike to work. We can encourage our kids to play outside, maybe even drop them off a few blocks from school so they get a little daylight and exercise before the first bell. Singapore and Australia have adopted recommendations that children spend at least two hours outdoors per day. In hopes of reducing high myopia rates, China has experimented with classrooms in which the walls and ceiling are made of clear plastic or glass to admit more daylight. Trends appear reversed in the US, where up to 40 percent of school districts have reduced or cut recess since the mid-2000s, partly in response to pressure from academic requirements associated with the No Child Left Behind Act.

Even with unfiltered windows and time outside, further factors can reduce how much light reaches our eyes. Smoke from wildfires, which are on the rise with a warming climate, suck out blue and violet photons. A little easier to sidestep is the possible overuse of sunglasses. Experts urge us not to wear that light filter every minute we spend outside. If the day is short, like it is during the winter in Seattle, doing so might eliminate almost any chance of receiving sufficient circadian-stimulating light. The rising popularity of blue-light-filtering glasses and coatings on prescription lenses, including some contact lenses, poses another obstacle. Tom Brady, a former NFL quarterback, has endorsed a high-end line of blue-light-blocking prescription glasses—reminiscent of star quarterback Jim McMahon as a brand-ambassador for BluBlockers. McMahon, who attributed his constant sunglass wearing to a childhood eye injury, was just one of many celebrities who created a cult following that persists today. The early 1990s infomercial with Dr. Geek's freestyle—"My name is Geek, I put 'em on as a shocker. . . . Man, I love these BluBlockers"—advertises the sunglasses as blocking 100 percent of ultraviolet and blue light. You could argue we didn't know better back in the early 1990s. The third photoreceptor had yet to be discovered; textbooks referenced only rods and cones. Today, however, it is evident that while blue light at night can be a problem, the same wavelengths of light remain essential for our health during the day.

FOR ANDREAS BILLMAN, THE *DAYLIGHT ROBBERY* PHOTOGRAPHER, the relic bricked-up windows had particular resonance during the pandemic lockdown. Both historical periods limited access to light and air. Both opened eyes to the pricelessness of both.

Many of us stayed indoors more than ever during the pandemic, and that's saying a lot given how much time we were already roofed and walled away from the sun. Newly minted home offices sprang up in converted basements, laundry rooms, even closets and attics. Any existing office building standards don't cover daylighting in those workspaces. Working from home also eliminated the commute. While people celebrated the recovered hours, they may not have recognized the lost doses of healthy light they had regularly received on those rides, walks, and even drives to work. At least for nine-to-fivers, commute times include those morning hours when the circadian system is primed to take in daylight.

Living and working in the same space can mean little contrast between day and night. There were exceptions. For some people, relaxed pandemic schedules allowed more time to take walks or exercise outdoors. The trail around Green Lake by my home became a highway of people. During my December 2021 stay in Manhattan, I noticed an unusual number of people sitting outside, sipping coffee, and basking in the sun—wherever the sun was accessible between the shadows, anyway.

The pandemic highlighted the importance of preserving sunlight in parks and open spaces. Central Park wasn't the only New York park losing sunbeams to new construction. Michael Kwartler consulted on the development of a new park in downtown Brooklyn, which will be in shadows nearly all day thanks to broad, tall buildings constructed at the southeast and west sides of the park. The buildings would have blocked less sunlight had they been located to the north, he told me. But Kwartler was hopeful that sunshine could still be preserved elsewhere, such as at 1960 Franklin Avenue in Crown Heights, where developers were proposing a pair of forty-story towers that would cast shadows on Jackie Robinson Playground and on the Brooklyn Botanic Garden. In San Francisco, a planning commission initially shot down plans for a proposed eight-story building in the Mission District that would frequently send a sizable shadow onto a neighboring school playground. Opponents warned that in the late spring and early fall, it would darken an estimated two thirds of the schoolyard during early morning recess. Ultimately, after years of lawsuits, developers still pushed the proposal through.

The pandemic has also highlighted opportunities to improve public health through thoughtful building design. More windows bring the dual benefits of more fresh air and more light. Both can fend off pathogens. As we threw open our windows to stop the spread of the virus, we also learned that the rays of sunlight beaming in could deactivate SARS-CoV-2 on surfaces. And, of course, sunlight sets and strengthens our clocks and, in turn, our immune systems.

The idea isn't novel. Florence Nightingale's revolutionary hospital design in the late nineteenth century involved large windows that allowed cross ventilation and filled rooms with natural light. "It is the unqualified result of all my experience with the sick, that second only to their need of fresh air is their need of light," she wrote. "After a close room, what hurts them most is a dark room. And that it is not only light but direct sun-light they want." Modern science supports her intuition: studies show that hospital patients recover faster and use less medication in rooms with daylight—and a view.

Lisa Heschong argued that in addition to the benefits of daylight and clean, pathogen-free air, dynamic and interesting views are also vital for happy, healthy, and productive occupants. The brain evolved to anticipate future conditions, she explained, so it's most sensitive to information that varies over time—whether that's fluttering leaves, shifting levels of blue wavelengths at dawn and dusk, or contracting day lengths. "It's trying to determine rates of changes so that it can make better predictions," said Heschong. More information, including orders of magnitude more photons, usually reaches our eyes when we look out a window compared with when we look the other way within our indoor space. Plus, interesting views motivate us to look out our windows more often.

Recognizing all these potential wins, scientists and architects are teaming up to champion a brighter future, one in which protecting the health of the planet doesn't require sacrificing human health. Heschong and Cincinnati Children's Richard Lang are together nudging the window glass industry to pay attention to the science and consider changing the way they do coatings. Lang noted that engineers are working on technical solutions that could meet both energy efficiency and biology goals. Meanwhile, a call is now out to include the protection of daylight access in the World Health Organization's Health for All policy and the United Nations' 2030 Agenda for Sustainable Development. Daylight and circadian health are common denominators for many of the stated goals and, experts argue, should be recognized as a human right. A transition away from fossil fuels—which fits the stated clean energy and climate action goals—may itself attract renewed attention toward the right to light. In many places, laws protect homeowners who invest in rooftop solar panels. And at least one lawsuit has already been filed over shadows cast upon those panels. In that case, a neighbor's redwood trees were culpable.

Also notably absent from the United Nations' 2030 agenda is addressing the epidemic spread of light pollution. We are not only underlighting our eyes during the day, but overlighting them at night.

6

Bright Nights

It's an impressive picture, beautiful even, when seen from high above. I watched mesmerized late one evening as my Apple TV screen saver took me on a nighttime flyover of Earth. Even without visible geological or political boundaries, I had no trouble tracing the continents from the strings of brightly lit coastal cities. More dense clusters of lights meandered inland, connecting other cities and towns and setting whole regions aglow. I could easily make out London, Shanghai, and Los Angeles.

In other images from space, I recognized my city. The I-5 corridor curving from Olympia through Seattle to Everett looked like a fat glowworm. Still, a few dark patches remained visible in my corner of the country. I set my sights on one of the inkiest expanses, a few hundred miles southeast in central Oregon. Then I invited my dad along on a quest to escape artificial light and maybe to reconnect with the near-infinite specks of starlight we once enjoyed gazing up at together.

A truly dark sky was a big ask. More than 99 percent of people in the US and Europe now live under light-polluted skies. Streetlights, headlights, porch lights, billboards, and more permeate our cities and suburbs, turning night into day. Reflections off asphalt, concrete, snow, and clouds amplify the radiance. Rural areas aren't immune. Years ago, as an environmental health reporter covering the expanding practice of hydraulic fracturing, I drove through the rolling Pennsylvania countryside and witnessed enormous flames atop sky-high torches, each burning off natural gas through flaring. These spots, too, stand out in satellite images from space.

It's become glaringly obvious that the proportion of our planet's surface with natural cycles of light and dark is shrinking. And it's shrinking fast. Based on samples of local skies taken by more than fifty thousand citizens worldwide, we now know that the average night sky brightened by nearly 10 percent per year between 2011 and 2022. That compounds to light pollution roughly doubling every eight years. As an author on that study pointed out, "If a baby is born today in a place where her parents can see 250 stars, by the time she is 18 she will only be able to see 100." That's assuming current trends continue. Thankfully, this problem could be far easier to reverse than other environmental crises. At least a third of all outdoor artificial light produced globally is wasted. One scientist told me he believes the figure is closer to half. What's more, reducing that energy waste could even put a dent in some of those intractable challenges like climate change.

But this reversal has yet to commence. Not only are there ever more people with ever more lights, but the type of lights we turn on is changing. Photos snapped by astronauts aboard the International Space Station over the last decade plus illustrate how cool blue-rich photons of LEDs are rapidly replacing the warm orange hues of traditional incandescent lamps and high-pressure sodium streetlights. That surge in short-wavelength light produces greater trouble for our clocks, not to mention increasingly unpleasant and potentially dangerous glare. In photos of London taken over the last decade, you can see a golden-yellow sprawl in 2012 turn into a blinding bluish-white spiderweb in 2020.

Our seven-hour drive in search of the dark took my dad and me into Oregon's high desert, more than one hundred fifty miles from the nearest major metropolitan area. There, we found a campsite at Prineville Reservoir State Park. Officials had recently made great efforts to preserve the park's dark, earning it a coveted designation in 2021 from DarkSky International. The US-based nonprofit recognizes public parks, reserves, and places across the world with the least amount of light pollution. I noticed how only red lights illuminated the fish-cleaning area next to the reservoir, and how the lighting in the bathrooms had motion sensors and remained dark when not in use. Yet akin to dodging man-made clocks underground, avoiding man-made lights proved difficult even there. We were car camping, after all. Intermittent beams from powerful LED headlights, headlamps, and flashlights of fellow visitors streamed into our campsite and our tents. Even the automatic lights inside our own car caught me off guard when I, ironically, went to fetch my blue-light-blocking glasses.

Once the night fully settled in, after campers grew quiet in their tents and the moon had set, the sky performed as advertised: pitch black and star studded. I sat in my REI chair, tilted my head back, and began picking out constellations. It was, as I had hoped, reminiscent of the speckled sky, framed by a canopy of evergreen trees, that my dad and I had gazed up at from the rooftop of my childhood house.

With more people losing sight of the heavens from home, many of us with the means are hitting the road in search of it. Dark-sky tourism is on the rise across the US. Occasionally, the dark sky will pay a return visit on its own—and remind us of what we're missing. When the power went out and the skies went black in Los Angeles after a 1994 earthquake, 911 received numerous calls reporting mysterious objects in the sky, including a "giant silvery cloud." It was the Milky Way.

LIFE EVOLVED FOR BILLIONS OF YEARS UNDER THE COVER OF DARKness at night. The only sources of light pollution confronting our early ancestors were generated by nature—predominantly moonlight, lightning strikes, and wildfires. Humans eventually added manmade flames and then proceeded to rachet up their manipulation of light. Today, our capacity to defy darkness has reached a radically higher level. The repercussions go well beyond losing sight of the Milky Way.

When the sun goes down, artificial outdoor and indoor lights extend our days and fool our physiology. Light shines into our eyes from oncoming traffic. It trespasses through our windows from the street and beams from home lighting fixtures and digital screens. And it all prevents the circadian prompts needed to wind down for sleep. As I stared at my Apple TV, wondering how people in Shanghai could possibly sleep, light from that television screen might well have been making sleep more elusive for me.

Around sundown, the shifting colors and decreasing levels of light detected by our ipRGCs should naturally trigger the body to ramp up its production of sleep-signaling melatonin. That message is now often diminished or delayed. A study in Australia found that nearly one of every two homes had enough light to halve the pineal gland's nightly secretion of melatonin. Homes with modern LEDs fared worse than those with incandescent lights. Other studies come to similar conclusions. The University of Washington's Horacio de la Iglesia and colleagues studied two Argentinian towns with different levels of available electricity and found that people in a town without access went to bed earlier and slept nearly an hour longer than those in a town that could light up the night to some degree. Compared to undergraduates at his university, people in the town without electricity slept about two hours longer. Of course, light's impact on our biology may be one of many reasons for the later nights.

The implications go further. The body leans on nighttime signals to unleash a cascade of important reactions, from drops in core body temperature and metabolism to a rise in leptin, an appetite-suppressing hormone. When darkness doesn't descend, lots can go wrong. Artificial light at night has been linked with depression, obesity, poor blood sugar control, reduced sperm quality, increased risk of preterm birth, greater susceptibilities to infectious diseases, and further undesirable health issues. Light at night appears to affect the composition and daily rhythms of the bacteria in our guts, which research shows could alter our own rhythms in return. Science even points to a correlation between the intensity of light at night and rates and severity of COVID-19 infections.

In a study of more than eighty-eight thousand older people, published as a preprint in late 2023, exposure to brighter light at night was linked with a higher risk of death over the next six years. The light doesn't have to be all that bright to cause trouble. In a separate study of older people, researchers linked low light levels at night—equivalent to leaving a television or hallway light on, or bedroom blinds open to streetlights—with higher rates of obesity, diabetes, and hypertension. Another study suggested we may not even consciously notice the effects. Participants who felt they had slept fine with a light on had higher sleeping heart rates and higher surges of morning insulin compared with when they slept with the lights off.

The radiating clusters of lights in those images from space resemble a malignant cancer, growing and spreading to satellite metastases. The comparison may be fitting. A study using astronauts' photos of Barcelona and Madrid linked a doubling of the risk of prostate cancer and about a 50 percent increased risk of breast cancer with greater exposures to light pollution of the blue variety. Another study in the US used satellite images and found up to a 55 percent greater incidence of thyroid cancer in residential areas where artificial lights shone the brightest. While relatively elevated, the cancers remained rare in both studies. Even for people exposed to the highest levels of light at night in the US study, the incidence of thyroid cancer was just 0.2 percent over about thirteen years. Plus, results from epidemiological studies like these can be confounded by multiple factors. The evidence so far is too limited to confirm blame.

Not all research on artificial light at night even finds negative effects. The International Commission on Non-Ionizing Radiation Protection issued a statement in 2024 calling for higher-quality studies to drill down on the genuine impacts. They highlighted long-standing limitations, including inconsistent methods of measuring light exposure and potential bias among participants aware of their exposure. Scientists equipped with new metrics and increasingly vast datasets are continuing to sort out these connections, in parallel with ongoing research into the pathways by which light seems capable of changing our clocks and our physiology. Even where proof of causation may be lacking, many argue that the evidence is strong enough to take a precautionary approach. "Light pollution is really like air pollution, which doesn't only affect your respiratory system," said Phyllis Zee, chief of sleep medicine at Northwestern University. "It gets into the rest of your body and affects your overall health."

No one faces more light at night, while being among the most vulnerable to its deleterious effects, than a hospitalized patient. Delirium is extremely common for patients in intensive care units, and has been linked with circadian disruption. A lack of light during the day and too much light at night could be significant contributors, alongside the presence of other sleep wreckers—from alarms sounding and doors clanking to round-the-clock awakenings for pokes, prods, and meals.

Melissa Knauert, a doctor who studies sleep and circadian rhythms at Yale School of Medicine, bemoaned the lighting patients endure in her ICU. One side of the hospital tower is adjacent to a helicopter pad; another faces a "very well-lit" ten-story parking garage; and another looks out at a local high school that keeps stadium lights on across the property all night. The fourth side of the tower faces the internal hospital courtyard, yet it holds the fewest beds. Then there are the televisions mounted in each room, which often stay on throughout the night, Knauert said. One day, while taking light readings in the unit, she was surprised to find that the light from the TVs became especially blue during commercials. And if all that light and noise wasn't enough to derail their clocks and sleep, patients are also surrounded by blinking LED lights and computer screens. Each room must have a computer by the bedside for electronic medical records. "That means there is a computer with a screen saver next to the patient," she said. "I have spent not a small amount of time training people to turn it away from the patient."

OUR CIRCADIAN SYSTEM IS FAR MORE SENSITIVE TO AN EQUIVALENT dose of light at night compared with the middle of the day. And that sensitivity increases as the evening progresses, per that phase response curve. But how much light in the late evening is enough to hinder the body's production of melatonin and meddle with its rhythms? A short stroll through a grocery store in search of ice cream? A quick scroll through social media on a smartphone?

The most formalized advice to date comes from the same consensus statement that prescribed a minimum for daytime light exposure. The scientists recommend keeping indoor lights below a melanopic EDI of 10 lux in the evening hours, especially within about three hours of bedtime. That maximum is still far higher than the light you'd be exposed to from a candle or campfire or full moon. It's plenty to read by. Yet 10 lux is below typical evening light levels in about half of homes. And it's significantly lower than other places we might find ourselves in the evening.

Supermarket lighting can be upward of 750 lux. "Such brightness produces an invigorating effect on the buyer," according to one retail lighting manufacturer. Since that revelation, I've tried to avoid late-night grocery runs. Meanwhile, after a little experimentation with a spectrometer, I learned that my smartphone screen can beam a melanopic EDI of 60 lux at my eyes. "I think we're going to look at people looking at screens at night, like in photos from 2021, just as we do people smoking in old photos," one scientist told me. "We're going to say, 'Oh man, how did they not know what that was doing to their body?'"

Part of the problem is that screens emit more blue light than we can consciously perceive. Not all devices are equally troublesome. Certain e-readers use front lighting that directs light toward the screen rather than the eyes. Use of blue-light-filters and apps can also reduce levels. But while melanopsin responds strongest to blue light, it still responds to longer wavelengths. It just takes a lot more photons of light at that orange-red end of the spectrum to trigger the same reaction. If you had a bluish-white LED and a red LED that is ten times brighter, the red LED would have a greater effect. This is why most experts urge us to focus on dimming light at night.

The recommended rules get stricter at bedtime: a maximum melanopic EDI of 1 lux. Our circadian system reacts to such low light levels late at night in part because our pupils become adapted to the dark and, therefore, more dilated. More light can get in. This is particularly true when we wake up in the middle of the night. But how many photons can really get in with just a few seconds of an open fridge? Or a late-night trip to the bathroom? Jamie Zeitzer, a sleep and circadian expert at Stanford University, offered a rough guideline: any light "brighter than moonlight and longer than lightning" could affect you and your clocks. Yikes.

Curious, I tested the light inside my fridge and detected nearly 500 melanopsin-stimulating lux. I'm sure it takes me longer to scan the shelves than it takes a bolt of lightning to strike. Spending a few minutes in a lit bathroom could affect rhythms and sleep, too, Zeitzer told me. A soft night-light instead of overhead or vanity lighting could reduce the impact. I'm now even considering a toilet disco light. Or, slightly less outlandish, a red toilet night-light.

As with their daytime melanopic EDI minimum of 250 lux, the consensus group's recommended nighttime maximums are rough average estimates. The ideal target will vary based on factors such as age, sex, and our light exposure earlier in the day. Older women with low estrogen levels, for instance, may experience greater disruption to their rhythms from light at night. The recommended levels could still be too bright for children, too. Research has found that preschool-age children are especially susceptible, with dim light in the evening significantly threatening their sleep. "If we start disturbing sleep and circadian rhythms in young children, that sets up a risk for development problems," said University of Colorado Boulder's Kenneth Wright. Of course, children get scared of the dark, and that fear itself can keep them awake at night—hence, night-lights. One possible compromise: a circadian scientist I met has created a red sleep lamp for young kids.

Bathing in abundant light during the day, as my dad and I did in the high desert of Oregon, can buffer the effects of nighttime light exposure. For this reason and others, experts express more worry about too little daytime light than too much light at night. Zeitzer added another dose of perspective. Sure, we could try to turn off all electricity at sundown and more or less turn back the clock to a preindustrial state, but that's not going to happen. Excessive focus on light's dark side could lead us to overlook the role of other sleep thieves like coffee after dinner or Instagram in bed. Someone might turn on their phone's blue light filter and think they are fine. "Then it's not the light from a screen that is keeping you up, it is the content," said Zeitzer. "If it's stressing you out, then I don't care if it's all red light and no blue light." On the other hand, if something is relaxing, even if it involves a little light exposure, the benefits might outweigh the risks. Cue the cute otter videos.

WILDLIFE ALSO SUFFERS THE CONSEQUENCES OF OUR ARTIFICIAL light. Across the animal kingdom, creatures can fail to forage, mistime their camouflage wardrobe change, lose their bearings, mate too early or too late, or give birth when food is in short supply. The confusion causes populations to dwindle and overall biodiversity to decline. Birds start their morning songs prematurely thanks to artificial lights, much like those Alaskan birds under the midnight sun. The same lights can lure them off course during their migrations. A creature doesn't even need eyes to experience the circadian impacts. Research suggests that low levels of light pollution—below the intensity of a full moon—can deceive oysters' clocks, throwing off when they open and close their shells.

Perhaps most menacing of all, light pollution appears to be an important and overlooked driver in the collapse of insect populations globally. Like a moth to a flame, many insects are fatally attracted to light. Some will fly into it and fry. Others will exhaust themselves after hours of circling. Still others will lose their way. Brighter nights disorient dung beetles in South Africa, who look to the Milky Way for direction on where to bury poop. Artificial lights can also wash out bioluminescent communications between fireflies. Love letters may get lost on their way to prospective sweethearts.

The impacts on insects have ripple effects across ecosystems, with significant consequences for us. We count on insects to recycle waste, control pests, and pollinate about a third of our food. Research shows that light pollution disturbs the circadian clocks and time-compensated sun compass of pollinating monarch butterflies. Moths and other nocturnal pollinators might be in even greater trouble; their night-adapted eyes can be extra sensitive to low levels of light. In a set of Swiss meadows, a study found that insect visits to plants at night dropped by nearly two thirds under artificial light compared with dark areas. Streetlights appeared to alter the number of flower visits by insects during the daytime, too, further suggesting that artificial light at night might indirectly affect the entire plant-pollinator ecosystem—including the daytime pollination by bees.

Scientists also warn of links between the diminished dark and increases in vector-borne diseases and seasonal allergies. A mosquito species that carries dengue and Zika appears to bite more when bathed in light at night. And urban light pollution could delay the winter dormancy of mosquitoes, which might mean they bite us longer into the fall. What's more, under artificial light at night, leaf buds were seen to open nine days earlier in the spring and delay changing color by about six days in the fall—on par with predicted impacts of climate change. The result for many of us may be more weeks of sniffles, sneezes, and watery eyes. And concerns extend to climate change itself. Artificial light at night in coastal areas could affect the uptake of carbon in the oceans. Countless marine creatures count on cycles of light and dark to time their daily migrations up and down the water column. As these animals descend after feeding near the surface, their respiration, excretion, and death at depth can sequester carbon on the ocean floor.

"We would do well by not ignoring the science, even if human health is all you care about," Travis Longcore, an environment and sustainability professor at the University of California, Los Angeles, told me. "And if you care about wildlife at all, and the environment at all, or the cultural heritage of being able to see the stars, then the things that you would do to help those will help humans, too. So, why not?"

THERE IS ONE TIME OF YEAR WHEN MUCH OF THE WORLD GETS EXTRA zealous with night lighting: the December holidays. While every homeowner may not go as far as stringing up twenty-five thousand lights like Chevy Chase in *National Lampoon's Christmas Vacation*, satellite images show that nighttime lights still shine up to 50 percent brighter during Christmas and New Year's in many US cities. The impacts are felt by creatures great and small.

Scientists at Texas A&M University in Kingsville studied the squirrel residents of their university campus and found the animals were more likely to forage at night and get eaten by predators during the weeks that bright white holiday lights decorated on-campus trees. One of the authors lamented to me that the university had yet to take their suggestion to "use fewer bulbs or switch from clear white LED bulbs to a colored LED bulb." Maybe we should all consider opting for traditional red and green Christmas lights as a safer option for wildlife and ourselves than the popular blue and white bulbs—which, admittedly, have always been my favorite.

When I flew to New York City for that December visit, I could see the shimmer of Manhattan from my window seat long before our approach to JFK Airport. Somewhere in that sea of lights below were the more than fifty thousand multicolored LEDs draped across the Rockefeller Center Christmas tree—plus a bright white Swarovski star. The growing glow reminded me of those satellite and space station images, now progressively zoomed in. The lights shone ever brighter up close. My taxi ride into "the City That Never Sleeps" brought me through a tunnel of luminous and animated billboards before fully immersing me amid blazing high-rises.

Just as with our light-deficient days, urban areas best showcase our overly lit nights. Those space snapshots illuminate the rapid transformation: Cities worldwide are becoming bigger and brighter. Towering skyscrapers cast both dark daytime shadows and a bright nighttime light that is hard to escape whether you're a human or hummingbird.

One night of my stay, a friend and I strolled through Times Square, shoulder to shoulder with the mobs of other tourists in town for the holidays. I opened my iPhone's light meter app and pointed it at the massive LED displays radiating every color of the rainbow—pink for T-Mobile, yellow for peanut M&M's, green for American Eagle, blue for Amazon Music. A few of my readings reached daylight levels, upward of 5,000 lux. As anyone who has been to Manhattan knows, day and night have minimal distinction within those blocks. Times Square is, you might say, timeless. I later read that city planners mandate a minimum luminosity for these billboards, read in LUTS, or Light Unit Times Square.

Back uptown, city lights proved unwelcome trespassers. My friend and her husband were battling an LED streetlight in front of their second-floor apartment. The husband had pleaded with the New York City Department of Transportation to shield the light so that it illuminated only the street and not their living room and bedrooms, including where their young daughter slept. One day, frustrated with the lack of response, he pried open the streetlamp and unscrewed a fuse—and then had second thoughts. He quickly screwed it back in. Given it was the holiday season, my friends also contended with intrusive light from a blue LED snowflake hanging from the lamppost. It's no wonder that the sky glow of New York City can reportedly be seen from more than one hundred miles away.

That Upper West Side streetlight might have been less obnoxious about a decade ago, when it would have been fitted with high-pressure sodium bulbs. This older standard gives off a warm yellowish glow closer to that of the moon. By the time of my visit, New York City was among municipalities across the US, and the world, that had upgraded to energy-efficient LEDs. Most of these lights emit heavy doses of bright blue light. The Department of Energy and the American Medical Association had been somewhat at odds over the race to install the new street lighting. In 2016, a year after President Barack Obama accelerated the DOE's push to retrofit streetlights, the AMA adopted guidance for communities—urging the selection of LEDs that minimize the harm to humans and the environment. They recommended "using the lowest emission of blue light possible." Unfortunately, by that point, many cities had already installed the standard, static blue-rich LED streetlights, and others seemed to miss or disregard the advice.

The Department of Energy forecasts that LEDs will make up at least four of every five lighting installations—indoors and outdoors—by 2035. A move to LEDs is not all bad, of course. The technology is increasingly affordable, uses up to 90 percent less energy, and lasts up to twenty-five times longer than traditional electric lights. Incandescent bulbs have now been effectively prohibited from store shelves—with few exceptions—in the US and Europe. They simply can't meet new energy-efficiency standards. And as regulators progressively raise the efficiency target, and as market pressures compound, compact fluorescent and high-pressure sodium bulbs are poised to become further relics of the past. In fact, the only technology capable of meeting the minimum of 120 lumens per watt proposed by President Joe Biden's administration are blue-enriched LEDs.

A trio of scientists laid the technological foundation for the LED upgrade in the 1990s. While red and green diodes already existed, producing a proper white light required blue diodes, too. The scientists' invention provided the missing ingredient, later earning them the 2014 Nobel Prize in Physics.

The innovation in blue diodes, and the subsequent rise in LED lighting, came many years before the trio of circadian scientists would win their Nobel Prize and spark greater recognition of circadian science outside academic circles. Consequently, early manufacturers of the lighting technology paid little mind to our third photoreceptors. While the new lights may have given our rhythms an inadvertent, modest boost of circadian-stimulating blue indoors during the daytime, the brighter and bluer photons beaming from our upgraded lamps, headlights, and streetlights threatened to throw our clocks off at night. And they still do. Steve Lockley, now part-time with the University of Surrey and a circadian researcher at Brigham and Women's Hospital and Harvard Medical School, bemoans that when city officials or building managers install LED lamps to save energy, they often install a light with the wrong spectrum. "They just need to choose the right LEDs," he said. "The costs are the same and there are many options available, so there's no excuse for getting it wrong anymore."

Technology is making it more feasible to produce LEDs that can be tuned to healthy hues and intensities. A few cities, Seattle included, have announced plans to roll out warmer white LEDs as current LED streetlights burn out. But as Seattle admitted in 2018, the process will take a long time. Because of the long life span of LEDs—much of their appeal—it could be another decade or more before those lights fade to yield another justifiable opportunity to invest in healthier and arguably more inviting lighting.

In the months leading up to the new lighting rules in 2023, one New Hampshire man reportedly spent about $1,700 to stockpile thousands of the soon-to-be-retired incandescent light bulbs. He detested the harsh light from LEDs and sought a lifetime supply of warmer incandescent lighting. We've been here before. Robert Louis Stevenson reportedly loved gaslights and hated the new electric lights that emerged in his time.

"A new sort of urban star now shines out nightly, horrible, unearthly, obnoxious to the human eye," he wrote. "A lamp for a nightmare!"

IN HIS CLASSIC BOOK *1984*, GEORGE ORWELL DEPICTED ROUND-THE-clock illumination as a torture technique. In 1990, a federal court in the state of Oregon deemed the same practice "unconstitutional" for a jailed prisoner. And, in 2018, while the US government held children in a Texas warehouse under constant overhead lighting, as part of the Trump administration's "zero tolerance" immigration policy, an editorial published in *The New England Journal of Medicine* warned of the "profound" consequences of that constant light, especially for their developing brains and bodies. "Ironically, immigrants coming from impoverished areas that may lack basics such as electricity may have left behind some environmental conditions—the natural light-dark cycle—that were far better for their health," wrote Brigham and Women's Charles Czeisler.

While these may be extreme cases, adults and children around the world are regularly deprived of darkness at night. Once again, those in low-income and minority communities bear the brunt. From blinding streetlights and headlights to the continuous glow of bars and gas stations and fracking flares, the onslaught of light at night is a class issue.

In Baltimore, just two miles separate a run-down section of Greenmount Avenue, brightly illuminated by LED streetlights and often a bonus mobile police floodlight, from charming cobblestone streets through wealthy, warmly lit neighborhoods. Samer Hattar, the circadian scientist with the US National Institute of Mental Health, lives in Baltimore. He worried that the lighting situation in low-income parts of the city perpetuated inequities—from inadequate sleep to unemployment. People in these neighborhoods will limit indoor lighting to cut electricity costs yet are flooded with streetlight at night. Especially after a weakly lit day, a disproportionate dose of bright light can wash out residents' rhythms.

The data back up Hattar's suspicions. People of color and low-income residents were found to be exposed to significantly greater light at night compared with affluent or white residents in both urban and rural areas in the US. On average, Asian, Hispanic, and Black Americans faced about twice as much light at night in their neighborhoods compared with white Americans. Another study broadly linked lower sleep quality to higher levels of night lighting, with the association stronger in poorer neighborhoods.

Disadvantaged families can face compounding hardships that brighten their nights. While bedrooms in wealthier urban homes might face away from the street, low-income residents' homes tend to be more crowded and use more rooms for sleeping, including those streetside, noted UCLA's Travis Longcore. Blackout blinds may be beyond the family's budget. And family members commonly share bedrooms; one sibling might be studying with a light on while another attempts to sleep. For some people, simply finding a safe space to sleep is the more significant challenge.

Robert Kwolek once lived in the Cowley Estate, a social housing complex in South London, where he could count upward of twenty lights out his bedroom window. Blackout curtains weren't enough to block the beams, he told me. The constant glare spilled in around the edges. "It would've been okay if that light had been a little warmer," he said. "But it was very blue." For the year he lived there, his sleep was abysmal. When he moved into another house where his bedroom faced away from the street, he slept far better. "Light pollution is the most underrated form of pollution," said Kwolek, now a senior architectural designer and project manager at Create Streets. The group is pushing for warmer-colored streetlights around London.

During my London wanderings, I witnessed how fewer and softer lights illuminated the wealthier neighborhoods. In Westminster, near my Airbnb, I basked in gentle yellow beams, including from a few remaining nineteenth-century gas lamps. At the time, controversy surrounded a local council's threat to replace the historic lights with LEDs. Less than four miles away, masted LEDs blasted down on stretches of Banner and Whitecross Streets, around another social housing complex. The stark illumination there resembled Kwolek's description of the lighting mounted above the streets and walkways, and on the sides of buildings, at his former complex: "It felt like I was walking in a prison yard," he told me. I felt a similarly oppressive vibe while walking the grounds of a couple of New York City public housing complexes. It made that blue LED snowflake shining into my friend's window relatively innocuous.

I MET MOHAMMED AS HE WAS SHOOTING HOOPS AT THE RIIS HOUSES, an NYC Housing Authority property in the East Village. Large LEDs mounted atop high poles lit up the blue and green paint of the outdoor court. It was almost 9:00 p.m., well after dusk. Standing at the free throw line, it could have easily been noon. Mohammed said he appreciated being able to play basketball into the night. He also told me he feels safer walking around after dark under the bright lights. But some of the lighting around the houses, he said, was a bit much.

New York City had recently begun changing out old, orange-hued lights on the property with the new bluish LEDs, as it already had with the streetlights around the city. I walked paved pathways through tunnels of lights between the residential towers, and through harshly lit parking lots. At one point, my gaze was drawn by LEDs and a rustling sound to a line of trash bags kept in perpetual motion by busy rats. As Mohammed pointed out to me, a few of the old lights remained mixed in with the new ones, highlighting the contrast in color and intensity. Most of the residents' windows that faced these new lights remained uncovered. Flags hung inside the glass on a few units. But I knew that such thin barriers were unlikely to serve well as blackout blinds.

Then there were the massive floodlights, the kind you might see on a roadway construction site. One had recently towered over this Riis courtyard before the police relocated it to another nearby housing complex. Mohammed recalled it running on a very noisy diesel generator and giving off a strong smell. I searched several blocks but never found the floodlight that night. A couple of weeks later, Mohammed texted me a photo of a floodlight that he spotted emblazing nearby Tompkins Square Park. Even through my phone, the glare of the tower's four huge spotlights almost hurt my eyes.

In 2014, New York City's then mayor Bill de Blasio launched a new crime-fighting strategy called Omnipresence. It would ultimately unleash hundreds of high-intensity mobile floodlights, accompanied by police presence, in low-income areas and public housing complexes across the city. The fleet still lights up the night, despite accusations that the program is the new stop and frisk and complaints from residents that the light sabotages their sleep. While erecting floodlights might be done with the goal of suppressing crime, the deployment may be inadvertently suppressing melatonin.

More thoughtful solutions should include "better—not brighter—lighting," and be "rooted in community engagement," according to leaders of a growing "Light Justice" movement. They argue that Omnipresence is just the latest in a long history of weaponizing light against people of color that dates back to before the invention of the light bulb. In the eighteenth century, "lantern laws" in New York City required Black people and other people of color to carry lanterns if they walked in the city after sunset and not in the company of a white person. Caught without one, they could be beaten or arrested. Those laws were followed by similar discriminatory policies. Arising in the late nineteenth century, and continuing into at least the 1960s, sundown laws excluded people of color from remaining in thousands of towns after sunset. In many communities today, underserved and marginalized residents still fear for their safety after the sun goes down.

RESEARCH ON WHETHER BRIGHTER LIGHTING MAKES A PLACE SAFER for anyone is inconclusive. The ambiguity regularly vexes decision-makers. As I write this, a debate is underway in Davis, California, over whether the city should amp up its street lighting in the wake of a series of stabbings.

More lighting would probably appease residents' fear of the dark, which some believe is an evolved trait. Scientists have shown that an absence of light can alter our neurobiology to stir up anxiety and fear. And, in certain circumstances, biological wiring that compels us to brighten our nights might still serve us well. One study found reductions in nighttime crime over three years of Omnipresence in New York City housing complexes. When the massive mobile LEDs were left on from sunset to sunrise, outdoor crimes around the light towers at night dropped. Yet the study's authors make clear that their findings were based on a tactical intervention in a specific setting and may not be generalizable elsewhere. Other studies of urban lighting reach conflicting conclusions. In Chicago, a rise in crimes reported to police followed a ratcheting up of lighting. In regions of England and Wales, rates of car break-ins dropped by half on streets that shut off lights for a portion of the night. Thefts might have been displaced to more brightly lit adjacent streets, according to the authors. And in China, incremental increases in nighttime light corresponded with increases in criminal arrest and prosecution rates. I'll add my anecdotal data point: my car was recently vandalized overnight in a public parking lot filled with an array of large LEDs.

Overall, the conclusion seems to be that perceived safety doesn't always translate to actual safety. Why wouldn't greater lighting lessen crime and accidents? A correspondence seemed logical to me at first. But bright lights can produce sharp contrasts between spaces where the photons reach and where they do not. This can give criminals deep, dark shadows in which to hide—between parked cars, for example. And the excessive illumination can not only create glare and blind spots, it can also help illuminate a target, whether an object or person. Glare can also blind people on roads and raise the risk of accidents, as anyone who has driven a vehicle since the surge of LED headlights knows.

At least one police officer discouraged the idea of more street lighting in Davis, suggesting it could backfire. "Crime requires light," he told local news. "Flashlights stick out like a sore thumb in the middle of the night."

AS THE SATELLITE IMAGES SHOW, INNUMERABLE NEW LIGHTS ARE switching on and turning up across the globe, even in the most remote regions. The rapid rise of off-grid lighting installations powered by solar adds momentum. In many places, the arrival of artificial lighting represents a welcome rise in wealth, security, and quality of life.

At the United Nations Climate Change Conference in Paris in 2015, the US joined with international partners in the Global Lighting Challenge, an effort to expedite the deployment of "high-efficiency, high-quality, and affordable lighting products, such as LEDs," across the world. The program aimed to reduce electricity use and carbon dioxide emissions, while creating "opportunities for prosperity for millions of individuals."

Light pollution isn't high on the list of concerns for developing countries. Unsurprisingly, many places simply choose the least expensive of these efficient lights—the mainstream white LEDs with their bright bluish glow. These LEDs now account for nearly half of global lighting sales.

While saving energy, pulling people out of poverty, and empowering kids to study into the evening are all very worthy aims, scientists urge more informed lighting choices that consider biology. The price gap between those basic white LEDs and the more tunable variety is shrinking. Michael Siminovitch, director of the California Lighting Technology Center at the UC Davis, is working with international partners to address the need for lighting standards. In Mexico, he said, people have picked cheap fixtures that shine excess blue light and dispatch the photons in all directions. He and other experts express concern that developing countries may miss their window of opportunity to opt for healthier alternatives. "It's a really serious problem because people are lighting like crazy," Siminovitch told me. What's more, he said, LEDs are so efficient that there is little motivation to ever turn off the lights. And leaving the lights on has long been a tough habit to break for all of us—even when the stakes have been more directly life or death.

DURING WORLD WAR II, KARL DÖNITZ, A GERMAN U-BOAT ADMIRAL who went on to succeed Adolf Hitler as leader of Nazi Germany, wrote of his surprise in finding US towns were "a blaze of bright lights." The urban glow was a gift for the Nazis, who proceeded to drown hundreds of Allied vessels and kill thousands of merchant sailors, military gunners, and civilians off the US East Coast. "For many weeks Nazi submarines have been sinking ships caught silhouetted against shore lights," reported *The New York Times* in April 1942.

Officials urged cities and citizens to shut off lights at night. But compliance remained low for months. In August 1942, the US Army issued a proclamation mandating blackouts and dimouts along the West Coast—up to 150 miles inland in some places. It banned nonessential lights and required shielding of essential lights. People began using vehicle headlight covers that narrowed and diffused beams. Black tape or cloth "of a grade comparable to flour sack material" over headlights reportedly also did the trick. That same month, a correspondent from San Francisco for *The New York Times* reported that "night baseball was played for the last time until the end of the war."

The East Coast lagged behind the West Coast in their mandates. Meanwhile, the British had beat all the Americans to it. To defend against air strikes during World War I, "lighting orders" compelled Londoners to turn off or dim unnecessary lights. People pulled old candlesticks and gas lamps out of retirement. A writer described the scene in November 1914: "Very few of the street lamps were burning, and these were so masked that their light fell only at their feet. . . . The town clocks are silent and at night their dials are not lighted." Britain enforced even stricter blackout regulations beginning two days before the country declared war on Germany in 1939. Civilians had been prepping for more than a year in advance by making blackout curtains and conducting regular blackout rehearsals, which helped pinpoint issues to resolve, like an illuminated clock in Ipswich's town hall that people couldn't figure out how to turn off.

Military technology evolved, and blackouts no longer proved passable protection during subsequent conflicts. But much of the world is now rediscovering the value of dark skies. In New Zealand, Indigenous Maori people are leading efforts to slash light pollution and earn the country certification from DarkSky. The Maori's practices and beliefs originate from their observations of the night sky. They measure time by the moon and count a day from sunset to sunset. Their fight to preserve the dark is also a fight against cultural genocide. If successful, New Zealand would become just the second Dark Sky Nation. The island nation of Niue won the first distinction in 2020.

More broadly, many countries, including France, Slovenia, and the Republic of Korea, have adopted strict national laws on when, where, or how people use lights. But the US is showing fewer signs of reducing its radiance. "We've had so much energy available," one source told me. "No one ever really thought about it." It sounded similar to the early arguments Americans made against restricting building heights. But there are a few bright spots—or, more apt, happy shadows. At least nineteen states, as well as the District of Columbia and Puerto Rico, have now enacted laws of some form that restrict light at night. Local campaigns to turn the lights out during bird migrations and sea turtle nesting seasons are on the rise, as is the availability of wildlife-friendly lights with longer wavelengths. "What's good for the creatures is more than likely good for us," said Siminovitch. Other grassroots efforts include the Soft Lights Foundation, which has petitioned the US Food and Drug Administration to regulate LED products and the National Highway Traffic Safety Administration, Congress, and the US Department of Transportation to ban "blinding headlights," citing a potential link with a recent rise in road fatalities. As I write this, I'm in the market for a new car, and I find myself steering away from another sedan toward a vehicle where I would sit up higher and hopefully above the direct oncoming beams. I'm even contemplating a splurge on one of those auto-dimming rearview mirrors.

The biggest push to lessen light pollution probably still comes from DarkSky, which recognized its first Dark Sky Place back in 2001: Flagstaff, Arizona. The city cashes in on the tourism boost. I noticed it even hosts Dark Sky Brewing, which, appropriately enough, brews an IPA called Circadian Rhythm. More than two hundred sites around the world are now DarkSky certified, including Prineville Reservoir State Park and the nearby resort community of Sunriver, Oregon, which my dad and I also visited on our camping trip.

Dark-sky advocates and scientists agree on a set of core tenets for minimizing light pollution: Light at night should be necessary, spatially targeted, temporally controlled, low level, and warm colored. Motion detectors, shields, timers, and dimmers can all help achieve these goals. Sunriver, for example, currently prohibits outdoor lights from facing upward or outward, and requires that they are shielded to prevent glare. Bob Grossfeld, who managed Sunriver's Oregon Observatory when my dad and I stopped in, called it a "win, win, win." He and others highlighted the copious benefits: energy and cost savings, dark skies for stargazing, a more natural environment for animals and plants, and stronger circadian rhythms for everyone. Well lit no longer equates to brightly lit.

I have come across innovative solutions, like glow-in-the-dark surfaces that absorb energy from the sun during the day and light up at night. I was fascinated by a project in Van Gogh Village in the Netherlands, famous for being the home of the Dutch painter. The village installed a motion-sensing system that lowers street lighting by 80 percent when there is no activity and then fully brightens when it detects a person—or bike or car—approaching. A circle of light surrounds the person throughout their journey. Comparable sensor-based systems for indoor spaces are rising in popularity. I sampled one with Siminovitch: as we walked down the hallway, lights just ahead of us sequentially popped on, and those behind us switched off—spotlighting us the whole way. A bonus of motion-detecting lighting: it calls attention to a criminal, much like a flashlight.

The substantial upfront costs required for such high-end installations are out of reach for many people. Not surprisingly, DarkSky has so far recognized a disproportionate number of affluent communities, like Sunriver. While artificial lighting may still be considered a sign of prosperity in parts of the world, the opposite is true in these wealthy areas—reflected by dips in light pollution. Here, nighttime darkness has become a desirable amenity.

The organization itself has called attention to the need for a more equitable dark-sky movement. Following dark-sky principles need not be technically challenging or expensive. Any community with the knowledge and will could put at least the basics in place and save money over time, experts told me. Meanwhile, despite the limited control we may have individually over our city's lightscape or our neighbor's brightly lit driveway or porch, we're not helpless as individuals. We can adjust the lighting on the inside and outside of our own homes to protect ourselves and our human and nonhuman neighbors. We can remember to turn off our lights at night. "A lot of people think you need a big huge light on your house to protect your belongings," said Grossfeld. "That's not really true."

Tools also exist to minimize the unsolicited light that creeps inside our homes and into our eyes—from blackout curtains to sleep masks to blue-light-blocking glasses. Yes, BluBlockers may have their time and place. And, yes, Corey Hart might have been onto something by wearing his sunglasses at night.

THE SAME SIMPLE TECHNOLOGY THAT WAS ONCE WIDELY USED BY British citizens to keep indoor light from getting outside could be leveraged more widely to prevent outdoor light from trespassing indoors. I spoke with one sleep scientist who proposed providing blackout blinds for people living near streetlights. That is the least we could do for people living in the beams of police floodlights. There is precedent: The British government provides grants to help soundproof the homes of people who live near noisy highways. And federal programs in the US allocate money to muffle the noise in neighborhoods near airports.

Just as during wartime, with each of these efforts, we must balance costs and benefits, necessities and practicalities, and the dangers of both too much light and too much dark. A 359-page report from the US War Department during World War II described a series of field tests that established the amount of sky glow produced by various types of lighting—vehicle headlamps, street lighting, illuminated show windows, traffic signals, building windows, outdoor commercial lighting, and industrial floodlighting—and determined the degree of dimming required to dampen sky glow and protect Allied ships. Engineers went so far as to specifically test the lighting hazards from Fords, Chevrolets, and Plymouths, as well as army vehicles and a thirty-passenger bus. The report also assessed the balance: At what point does an increase in driving risks outweigh any additional benefits with muting sky glow?

In October 1943, *The New York Times* reported on one lighting restriction that may have gone too far: red and green traffic lights had been reduced to a "thin cross difficult to see at night and still more difficult to see in daytime." The newspaper speculated that the faint traffic lights may have caused a rise in accidents. Meanwhile, across the United Kingdom, initial worries about a dramatic blackout crime wave turned out to be somewhat overblown, except for petty theft—and at least one serial killer. Fears about road fatalities did prove warranted. In 1940, at the peak, one person died for every two hundred vehicles on the road. The rate had nearly doubled compared to before the war. Posters cautioned the public: "Look out in the blackout. Until your eyes get used to the darkness, take it easy." White paint was soon applied as a safety measure, stripes decorating everything from street curbs to cows.

Another familiar concern crept up: a lack of light and air. Heavy curtains, black paint, and cardboard covered windows, darkening both night and day. They might as well have been bricked up. But there were bright sides to all the darkness. Blackouts created nighttime environments more conductive to sleep and supportive of circadian rhythms. They also unveiled the starry night sky. Radio broadcasts encouraged people to go outside and look up.

Technology has come far since World War II. The surge of LEDs poses multiple threats but, used optimally, also has great potential to create safe, healthy, and attractive indoor and outdoor lighting. New LEDs are more mutable and tunable than previous technologies and could help us ease the public health and environmental problems that indiscriminate use of the technology initially exacerbated. Even more hopeful, once we turn off light pollution, it disappears. It doesn't linger in the environment like industrial chemicals and other toxic contaminants. And unlike the burning of fossil fuels, which will continue to wreak havoc on our climate for years even if we completely ceased the practice, we could instantly bring back the dark.

My dad and I enjoyed our reprieve from light pollution while camping. Based on my hourly harvests of spit, which I collected in test tubes during the nights before and after the trip, I discovered that the bright days and dark nights had managed to bump up my nighttime rise in melatonin about an hour earlier. My dad and I also escaped the more recognized pollutants contaminating the air and warming the planet, generally spewed from automobiles, airplanes, and local industry. But just as we were packing up to leave on our final morning, a brown haze rose from the horizon. Wildfires had sparked across the region, and smoke was wafting our way. Minutes later, as we loaded the car, we began to smell it. Our drive back to Seattle took us through miles of burnt landscapes and thick smoke.

Wildfires have become an increasingly regular occurrence as the climate continues to change. Much like cloud cover, a layer of smoke—or any air pollution for that matter—can increase how much artificial light is reflected back to the Earth's surface at night while decreasing the sunlight that reaches the surface during the day. On top of that, particulate matter itself is on a growing list of clock scramblers in our industrialized world—from pollutants we struggle to keep out of our bodies to foods and beverages we intentionally take in.

7

Clock Scramblers

The recipe called for banded mystery snails, bridal shiners, fingernail clams, pill bugs, and minnows, along with a hefty helping of algae, zooplankton, sand, and leaves. And it included a key ingredient in any American recipe: salt. This soupy mix was obviously never intended for the dinner table. And the salt wasn't that fine snow-white stuff found in most table saltshakers either; it was the kind widely dumped on streets and sidewalks when temperatures drop and snow falls.

Kayla Coldsnow and her fellow environmental biologists at Rensselaer Polytechnic Institute in New York knew road salt doesn't stay put. Eventually, as temperatures rise and snow melts, much of it gets carried into the soil and then into neighboring lakes and streams, where it can disrupt freshwater ecosystems. Commonly at the base of those living systems is a type of zooplankton called *Daphnia*. It is another one of those tiny organisms that spends its days and nights commuting up and down the water column. Their journeys are part of what is widely heralded as the largest daily migration on the planet—a massive wave of life that ripples east to west with the sun, across every pond, lake, and ocean. A lot is riding on it.

Coldsnow and her colleagues wanted to know whether higher salt levels in the environment might be affecting *Daphnia* and, therefore, causing trouble all the way up the food chain. To investigate, Coldsnow's colleagues put dozens of three-hundred-gallon cattle tanks in a field, filling each with the fixings of a lake—plus varying amounts of salt. Then they waited and watched.

Initially, they saw *Daphnia* suffer enormous die-offs when exposed to salty waters. But then, somehow, their populations bounced back. It seemed the short-lived critters could rapidly evolve a tolerance to cope with excess salt. However, that adaptation appeared to carry a trade-off, as evidenced by the zooplankton's subsequently slow population growth. What had *Daphnia* given up to survive the onslaught of salinity? What were the long-term implications for its population and the other species that counted on it?

The search for these answers led Coldsnow and her doctoral adviser, Rick Relyea, to team up with a circadian scientist at the institute, Jennifer Hurley. They were curious about whether *Daphnia*'s clocks might have been compromised by the salt. Hurley herself was initially doubtful. Her reasoning leaned on one of the fundamental principles of circadian biology: clocks should be buffered against environmental insults, including those notorious fluxes in temperature—and salinity. But, sure, she'd help them take a look.

To Hurley's surprise, experiments in her lab showed that the salt exposure did in fact mess with *Daphnia*'s clock. The saltier the water, the weaker the ticking. At high salinity levels, oscillations in clock gene expression completely stopped. *Daphnia* essentially lost track of time.

What happened in those cattle tanks isn't bad news for only *Daphnia* and the ecosystems that depend on it. The genes and processes involved in *Daphnia*'s clockworks are very similar to those that regulate our clocks. While humans might not be exposed to comparably lethal levels of salt, we are increasingly exposed to a host of other clock destroyers, both in our environment and in our diet.

MANY TRADITIONAL CONTAMINANTS ARE EMERGING AS PROBABLE circadian and sleep disruptors—flame retardants, heavy metals, particulate matter, ozone pollution, cigarette smoke, PCBs, solvents, and cyanobacterial toxins, to name a few. In one study, exposure in the womb to small amounts of bisphenol A—an estrogen-like synthetic substance added to plastic containers, food can liners, and paper receipts commonly referred to as BPA—exacted a lasting toll on mouse clocks that was passed on across generations.

The timing of these exposures likely influences the degree of harm they inflict on our bodies and our timekeepers. Ingesting the same leaching bit of BPA or breathing the same dirty air in the morning could have a different impact than doing the same at night. "It looks like the clock is fine-tuned to be sensitive to the environment in ways we never thought," said Hurley.

Much of my journalistic work has focused on toxic exposures in our environments. It's been eye-opening and extremely depressing to learn how ubiquitous and insidious these threats are in our everyday lives. Pollution is deemed responsible for one in six deaths worldwide—on par with deaths caused by smoking. Still, despite more than a decade of reporting in this area, I was stunned to hear the suggestion that a good portion of the known health effects of environmental exposures may result from a pollutant's interference with circadian rhythms. Because our clocks help regulate so many physiological functions—from metabolism to the immune response to reproduction—damaged clocks can inflict wide-ranging damage on the body.

Exposures to salt and other contaminants might not immediately kill us, Coldsnow explained, but could produce subtle yet consequential effects that we're only beginning to understand. She and Hurley speculate that we evolved to divert energy away from physiological functions that might not be absolutely and acutely necessary—like keeping internal clocks ticking in sync—when we need to address more urgent matters such as surviving a toxic exposure that we can't escape.

Many more pervasive pollutants, personal care products, and pills get into our bodies and warp our clocks. The widely used drugs metformin and rapamycin, as well as certain anti-tumor treatments, beta blockers, and antidepressants, are all known to affect our rhythms directly or indirectly. The same may be true for oral contraceptives, which hundreds of millions of women around the world take every day. Hurley expressed her bewilderment that the impacts remain largely unexplored despite the known effects of estrogen on circadian clocks. One study from 1996 suggested oral contraceptives "obliterated the circadian rhythm of melatonin," in addition to upsetting daily patterns of cortisol, blood pressure, and heart rate. More clock scramblers infiltrate via the air and water. Pesticides can hitch rides on our food. And then there are the hazards that we ingest wittingly—like bottomless sodas and heaping helpings of salt. "We put things in our bodies long term that we never would have done one hundred fifty years ago, fifty years ago even," said Hurley. "We don't know the half of what we're doing to our circadian clocks."

SOMEWHAT LESS SURPRISING TO FIND ON THE LINEUP OF CLOCK wreckers are caffeine and alcohol. The two drugs bookend the day for a lot of people. Both can make it harder for the body to decipher the day's beginning and end.

As Michael Pollan explains in his book *This Is Your Mind on Plants*, caffeine deserves much credit for regularizing the modern day. "The power of caffeine to keep us awake and alert," he writes, "to stem the natural tide of exhaustion, freed us from the circadian rhythms of our biology and so, along with the advent of artificial light, opened the frontier of night to the possibilities of work." Caffeine's power also locks eyes wide-open to stare at ceilings late into the night.

Caffeine affects both the circadian and sleep homeostat systems, and can linger far longer than we may realize. On average, about half remains in the body after five hours—possibly a lot more of it if you are a woman taking oral contraceptives, which can slow caffeine's clearance. Some of your venti cappuccino could still be in your bloodstream twelve hours later, hijacking adenosine receptors and disrupting both sleep quantity and quality. Research has even shown that caffeine may alter the circadian system's response to light. With most Americans drinking coffee every day, that's a lot of opportunity for confused clocks and sacrificed sleep.

Clearly, we should be limiting our afternoon or evening caffeine buzz. Yet seeking a fix first thing in the morning carries potential consequences, too. Doing so could interrupt both the body's morning cortisol pulse and its clearance of nighttime adenosine. Payment for messing with these delicately balanced rhythms might include a worse afternoon crash. If you must have that cup of morning Joe, some experts recommend a sweet spot: between 90 and 120 minutes after waking.

I disregarded the advice at first. The only thing getting me out of bed many mornings is the vision of my first pour-over. But my midday slumps were rough. So I reluctantly gave the recommendation a try, holding off on my coffee for about an hour and a half after crawling out of bed. While I missed my morning ritual, I noticed that I still had energy to start the day. Placebo effect or not, I also noticed that I didn't slump quite as low in the afternoons. Of course, every body is different in how it processes and responds to caffeine. In one study, morning larks awoke more during the night than night owls after consuming coffee during the day. You might need to do your own experimentation.

Then there's alcohol, which usually comes into play later in the day—with the exception of Bloody Marys and mimosas. Aside from the hangovers I've experienced over the years, I had never realized the extent of alcohol's impact on my sleep. A glass of wine always seemed to help me unwind and relax before bed. But as soon I began paying attention, with the help of my Fitbit, the correlation was clear: Even after just one drink, especially if it was close to bedtime, my heart rate would remain relatively elevated through most of the night. And my sleep score in the morning would match my energy and mood: painfully low.

Alcohol can initially help us fall asleep. However, research shows that drinking more than two alcoholic drinks per day for men or more than one per day for women can decrease resting heart rate variability, a proxy measure for sleep quality, by nearly 40 percent. Rapid eye movement, or REM, sleep is especially at risk. Scientists also liken a hangover to jet lag: alcohol can uncouple the liver clock and master clock in the brain, as well as obstruct the circadian release of hormones and the rise and fall of core body temperature. Alcohol may suppress the production of melatonin. And it can further impede sleep by raising the risk of sleep apnea. If you do want to enjoy a drink, experts advise doing so at least a few hours before bedtime.

Of course, you and I probably don't reach for a cup of French press coffee or pint of beer with any notion that we're doing our bodies good. The same goes for french fries or a pint of ice cream. Public health organizations continue to call us out for overconsuming salty, fatty, and sugary foods, which we tend to do much more intentionally than *Daphnia*. We've been warned about the repercussions of these dietary choices our whole lives: high blood pressure, obesity, diabetes, heart disease, cancer. Now we can add perturbed rhythms to that list. Studies have linked a high-salt diet with disrupted sleep and circadian timing in mice. The same may be true for us. Research has also shown that the circadian clocks in animals who eat a lot of fat can break, resulting in obesity for many. Too much sugar can meddle with sleep and the circadian system as well. An overnight spike in blood sugar is almost a given. As other research cautions, eating a lot of saturated fat and sugar, in addition to too little fiber, can mean lighter, less restorative sleep and more arousals.

Yet what do many of us do? We raid the freezer for a late-night bowl of ice cream. And if we find that freezer bare, the cookie jar empty, and our brightly lit neighborhood grocery store closed, we probably still have options.

IT WAS 11:00 P.M. WHEN I TAPPED OPEN THE GRUBHUB APP ON MY iPhone. That move alone broke the rules of proper circadian hygiene. Then I started to scroll. What tasty treats could still be delivered to my door at that hour? Opportunities to satiate just about any late-night craving were available: shawarma, pizza, burgers, tacos, ice cream, cookies. The aptly named Midnight Cookie Co. offered Red Bull, s'mores cookies, even vegan ice cream. Several competitors were also willing to bring me dessert—from Pie Bar to Never Too Sweet. All appeared to be following the crumb trail of Insomnia Cookies. That expanding US chain was started out of a University of Pennsylvania dorm room in the early 2000s to fulfill college students' desires for late-night sweets. Now, it delivers "warm cookies until 3:00 a.m. daily" across multiple zip codes.

As tempting as a midnight cookie or pint of ice cream sounded, I'm proud to report that I stopped short of placing an order that night. Sure, I could've excused it as "book research." But I decided that I'd wrecked my clocks and sleep enough over recent months. I am no Morgan Spurlock.

While I might have been overly aware, Midnight Cookie Co.'s typical blissed-out customers are probably not thinking about the consequences of the gobs of fat and sugar on their internal clocks. And it's fair to guess that only a rare customer knows to consider another critical factor: it's not just what or how much you eat but also *when* you eat that matters.

Half of adults in the US eat, snack, and nibble across at least fourteen hours each day—nearly from the moment we open our eyes in the morning to when we close them at night. Only around one in ten consistently eats within a twelve-hour stretch. That wide window doesn't look to be shrinking, at least if the food industry has its way. In 2023, Post Consumer Brands began selling a Sweet Dreams cereal, which it purports will "support a good sleep routine and a fresh start to the next day." The product comes in two flavors: Blueberry Midnight and Honey Moonglow. Both are chock full of sugar—sixteen grams and thirteen grams, respectively. Post appears to be among companies trying to compensate for recent declines in the sales of cold breakfast cereals, critics say, by creating a "fourth meal" of the day. And if boxed cereal is not your thing, you can now splurge on some fourth-meal Sleepy Chocolate from The Functional Chocolate Company.

It's not just Americans who eat over extended hours of the day and night. Studies of adults in Switzerland, Australia, and India have found similar patterns. This is all at odds with how our ancestors ate—usually in very compressed periods during daylight hours, with long periods of fasting overnight. It's fair to assume that hunter-gatherers rarely risked their lives to search for a midnight snack. Even after the Agricultural Revolution, as people enjoyed more reliable access to food, they still generally consumed it between sunrise and sunset. The ancient Greeks typically ate two meals a day: breakfast and lunch. During the Middle Ages, a third meal grew more common. But that supper was small and finished before twilight. The Industrial Revolution, with its extended working hours, long commutes, and greater availability of light, pushed the last meal later. And it grew larger. Changing cultural norms and emerging technologies continued to loosen the link between cycles of light and dark and feasting and fasting. Today, the convenience of processed foods, refrigerators, microwaves, twenty-four-hour restaurants, and food delivery apps has all but eliminated the connection.

OUR BODIES EVOLVED TO SEEK FOOD, EAT FOOD, DIGEST FOOD, AND eliminate excess food at certain times of the day. Ghrelin, the hunger hormone, is released from the stomach in a circadian rhythm. So are the enzymes needed to break down food. While certain rhythms peak during the day, others ascend at night. The pancreas slows its drip of glucose-regulating insulin in the evening—a tap further clamped by the coincident rise of melatonin secreted from the brain's pineal gland. Come morning, if all is ticking right, most of us can expect our motility rhythms to compel a bowel movement. But when we don't abide by our natural program, when we eat erratically and across a wide window, we perplex our clocks and cause everything from that critical prep to the cleanup to go undone. Incoming food can then injure the body much like a breath of polluted air or a dose of a toxic chemical. "A meal is a physiological insult," one scientist told me. "You have to be prepared for it."

One of my first lessons in this came during a conversation with Martha Merrow, a chronobiologist at LMU Munich. "I haven't eaten my dinner yet," she told me. "This is like the worst thing—you're not supposed to eat at ten o'clock at night." I felt bad. I was the one who suggested we chat over Zoom at that hour, late enough in my time zone that I didn't have to revise my schedule and bungle *my* body clock.

As we've learned, the master clock in the brain is primarily entrained to the twenty-four-hour day by exposure to light and dark. And it forwards that intel to enlighten clocks throughout the body. But peripheral clocks are "promiscuous," as one scientist put it. They can favor other cues, such as incoming calories. Consequently, late meals can uncouple peripheral clocks from the master clock. Evidence suggests they can even directly alter the ticking of the master clock itself. By eating so late, Merrow told me, she would miss her body's anticipated eating window and cause the circadian phase of her liver to shift "very, very, very late." That could have repercussions near and far within her body. Her kidneys would surely miss the mealtime memo from her liver because it arrived so late. They might then lose track of time and fail to suppress urine production as they normally do while we sleep. "And maybe I'll be getting up all night," she said. Okay, now I really felt bad. I would also later realize that a comparable desynchrony probably explained why I got up to pee four times after my pancake "dinner" in the bunker.

The problems go beyond late-night trips to the bathroom. Core body temperature naturally dips at night. An internal cooldown of a degree or two Fahrenheit promotes sleep. Eating late derails that rhythm. Your body responds to food, or a caloric drink, by sending blood rushing to your gut, which raises your core body temperature. It can take upward of two or three hours for the stomach to empty after eating, even longer after larger or fattier meals. And let's be honest, if you're eating late at night, you're probably not eating a salad or bowl of soup.

To make matters worse, a lack of sleep has been shown to trigger an excess of ghrelin and to promote poor food choices, like ordering a dozen cookies for yourself. Placing a cookie order could even start a vicious cycle. A high-fat diet has led mice to eat more during daylight hours, the equivalent of our nighttime. If this finding translates to humans, eating more cookies might drive you to eat yet more cookies late at night. And that might in turn increase your desire for more unhealthy snacks the next day and night.

Unfortunately, even if your late-night snack was low in fat, salt, and sugar, you're not off the hook; your clocks could still fly awry. Our bodies are simply less primed at night to digest and metabolize food. "Your liver at midnight, your liver at noon, are totally different organs," said Russell Van Gelder, the ophthalmologist at the University of Washington. Depending on the time of day or night, the liver will respond to the same food differently. For example, insulin promotes the uptake of glucose in the liver for later use and, more generally, lowers blood sugar levels. Because insulin secretion and the body's sensitivity to insulin are lower at night, glucose levels will rise higher in response to an equivalent calorie. On my second night in the bunker, before my clocks lost the beat, I ate a big but relatively low-sugar dinner of chili—and, yes, a few chips—at about 8:00 p.m. Data from the glucose monitor showed my blood sugar levels were still rising as I got ready for bed a couple of hours later and that they then plateaued while I lay awake over the next hour or two. Our fat-burning system also doesn't rev up until several hours after we stop putting fuel into our mouths. For many reasons, later calories are more likely to add pounds or body fat.

It's indisputable that veering from our ancestors' eating and drinking patterns is throwing clocks out of whack. However, the science does point to a glass-half-full interpretation: consolidating the hours during which we consume our calories might be an easy and powerful way to right our clocks and improve our health—on par with switching off lights at night. A growing area of circadian research is focused on so-called time-restricted eating.

UNLESS IT'S A SPECIAL OCCASION—SAY, A FRIEND IS GRILLING UP A pig or pouring glasses of wine late into the night—Victor Zhang avoids consuming calories before 10:00 a.m. and after 4:00 p.m. He's an avid boxer. He told me he hasn't noticed any drop in his strength or energy levels after restricting his eating times. If anything, his performance has improved. I also got the impression that he worries little about the composition of the calories during his shortened eating window. At one point during our meeting at the University of Washington, I watched him dash off to partake of a batch of brownies one of his lab colleagues had baked. Zhang, a postdoc in circadian science when we met, said it isn't hard for him to "eat at the same time like a laboratory animal." And he highlighted a bonus benefit: "It also makes my life more convenient, since I can spend less effort planning my day around the next foraging opportunity."

After several conversations with circadian scientists who shared how they had shrunk their daily eating windows to ten hours, eight hours, even a mere six hours for Zhang, I began to think there might be something to this time-restricted eating thing. I already knew that religions and cultures worldwide have been practicing forms of fasting for millennia. And I had witnessed the rising popularity of intermittent fasting in recent years. So how did this diet strategy differ?

While there is overlap, time-restricted eating is done consistently—the same range of hours every day—and doesn't try to restrict calories.

It's easy to imagine how a longer period of fasting might naturally result in a person consuming fewer calories and, thereby, shedding a few pounds. Yet some studies show weight loss even when calorie intake is equivalent. And what has scientists most excited are the potential impacts on disease.

In a study published in 2012, scientists from the Salk Institute for Biological Studies fed two groups of mice the same high-fat diet known to be linked to chronic metabolic conditions like diabetes and heart disease. They allowed one group to eat it across the twenty-four-hour day and the other they restricted to eating the same food across only eight hours each day. Mice who ate those unhealthy calories in the tighter eight-hour window appeared to be protected from the diseases.

In 2019, researchers from Salk and the University of California, San Diego, strengthened the case by finding that a small group of people with metabolic syndromes who limited their calories to a ten-hour window, in combination with their medication, saw significant improvements in body composition and cholesterol levels. Other studies in animals and humans more carefully controlled caloric intake and reached similar conclusions. The list of proposed benefits now extends to include improved cancer survival, reduced inflammation, enhanced endurance, and a bolstered blood-brain barrier. In studies of mice, limiting the eating window improved the sleep and cognitive deficits, as well as reduced the beta-amyloid plaques, associated with Alzheimer's disease. Maybe most intriguing of all, scientists see hints of a connection between time-restricted eating and longevity.

The notion that reducing calories could extend animal life span has been floating around for more than eighty years. But the research that spawned that idea restricted calorie intake while not paying attention to when during the day the animals ate. More recent studies found that animals largely changed their feeding pattern when given a limited-calorie diet: they ate all their food for the day within a few hours and followed that feasting with a long period of fasting. That begged the question:

Could timely fasting explain some of that resulting extra life time? Joseph Takahashi, a neuroscientist at the University of Texas Southwestern, and his colleagues tried to find an answer.

After designing and constructing feeders that automatically dispensed food pellets at preprogrammed times—no one in the lab wanted to sacrifice their own circadian clocks to hand-feed mice 24-7—they found that a reduced diet gave mice an approximate 10 percent boost in life span. That was impressive. But even more remarkable were the mice that ate the same limited chow constrained to a twelve-hour window. If that window fell at their wrong circadian time—daytime—the mice lived about 20 percent longer. If it fell at their correct circadian time—nighttime—the mice lived about 35 percent longer.

The team found no significant benefit to restricting the eating to a narrower two-hour window nor any impact of timing on body weight. That means, Takahashi told me, the effect on longevity was probably independent of any effect on weight. If the same is true for humans, it would mean that we, too, could live longer by watching when, not just what, we eat. To prove it translates is an incredibly challenging task, however. Longevity in a human is measured on a far longer timescale than in a mouse. And controlling what, when, and how much a person eats over several decades is hardly practical. Takahashi suggested that researchers could at least collect clues by measuring and comparing the inflammatory markers associated with aging in people. "Aging is really a disease of inflammation," he said.

In their study, Takahashi and his team noticed that the non-fasting mice had far more changes in gene activity as they aged, including changes related to inflammation, than mice in the other groups. Calorie restriction appeared to go a long way in rescuing the youthful gene function, but circadian alignment of feeding time added "another level of protection," he said. Further studies add to the evidence. Salk scientists found that restricting the eating window for mice to between eight and ten hours decreased the activity of genes involved in inflammatory signaling.

Other research teams are investigating the potential longevity link, including Michael Young and his colleagues at Rockefeller University who found that fruit flies live up to 50 percent longer when their food was limited to twelve hours a day. Results so far have overthrown the paradigm that a "calorie is just a calorie," according to a 2022 National Institutes of Health workshop. While the leading experts in attendance concluded that the science does not yet guarantee healthier aging or enhanced longevity for humans who limit their eating to shortened windows, hope and enthusiasm are high. Time-restricted eating is "like a time giver," said one scientist in Young's lab, the translation of *zeitgeber* not escaping her. "Hopefully we'll be able to figure out what the exact mechanism is, and then maybe we can target that without any of the diet. You know, like the Holy Grail."

WITH SUCH SEEMINGLY PROFOUND IMPLICATIONS, I WANTED ALL THE details. For starters, what actually counts toward that break in the fast? Does a calorie or two do it? How about coffee with a tiny splash of cream and sugar? Or black coffee? The answers to these questions, I learned, are heavily debated. Part of the problem is that much of the research has been on rodents, Emily Manoogian, a staff scientist at Salk, told me. Those studies generally randomize the animals into only two situations: they are given either standard, boring rodent cuisine and water, or only water. "Rodents don't get coffee," said Manoogian. As the science has expanded to clinical trials in humans, cost pressures force researchers to control as many factors as possible to reduce how many people they need in the trial. This is why her research defines the "break" as consuming anything other than water. Manoogian admitted that the science hasn't yet come to a verdict on things like caffeinated coffee.

Okay, so the eating clock starts with the first bite of a bagel or sip of a latte and stops with the last crunch of a potato chip or splash of wine—

or, ideally, healthier choices. But how tight does that eating window really need to be? And does it matter if it's earlier or later in the day?

Many studies have been quite strict—limiting meals to six or fewer hours and cutting people off from calories after midafternoon. Thankfully, we don't need to go that extreme. Manoogian's studies usually ask volunteers to aim for a ten-hour eating window, which she suggested is a sweet spot for balancing benefits with tolerability. It's usually best to hold off on eating until at least an hour after waking up, she said, and then stop at least three hours before bedtime. Life can get in the way, of course. "I have a daughter and the only time I get to exercise is after I put her to bed, before I eat," said Manoogian. "So, my dinner is a little bit later than I would like it to be."

Although less compatible with typical social calendars, concentrating calories earlier as opposed to later in the day seems to reap the most rewards. The science appears to align with the popular adage, "eat like a king in the morning, a prince at noon, and a peasant at dinner." Our intuition is being confirmed yet again by modern science. One study compared a group of overweight and obese women eating according to a plan reflecting this advice with a similar group following the reverse and more typical American meal pattern: a small breakfast, a medium-size lunch, and a large dinner. Over the course of a day, the women all consumed the same calories. By the end of the study, the women who ate like queens earlier in the day had lower blood sugar levels and lost more weight compared with the other group. Further studies underscore the potential health and weight loss benefits of earlier eating coupled with longer nighttime fasting, as opposed to skipping breakfast. Still, more long-term large-scale studies of humans are needed to determine optimal eating patterns. And the precise prescription will always be personal.

Important contributors to healthy digestion peak after we wake up in the morning. Insulin sensitivity was found in one study to be 54 percent higher just before noon compared with midnight. For most of us, late morning through early afternoon is when our body is most prepared to handle and break down, say, blueberry pancakes and use the resulting fuel. Eating during an earlier window also reduces evening levels of cortisol and ghrelin. Meanwhile, restricting food intake to intervals in the late afternoon or evening appears to have no benefit or may even worsen health measures, such as blood sugar and blood pressure. But this doesn't mean you should down that bowl of cereal first thing. As with coffee, Manoogian and others still recommend waiting a little while to allow your melatonin levels to drop and insulin secretion and sensitivity to rise. While the weekend brunch may have earned its popularity coincident with the rise of social jet lag—we had to sleep in on Saturday and Sunday to make up for those early alarms on the other days—I can now see its multiple merits.

Scientists generally agree that dinner should be early, light, and lean more on proteins and fats than carbohydrates. Again, those daily insulin rhythms mean the body is less able to handle carbs later in the day. The circadian rhythm of saliva production is another sign that we weren't meant to eat late. Far less is secreted at night to aid in the early stages of digestion. This may also help explain why we tend to wake up with dry mouths.

The jury is still out on whether it matters how many meals you eat during the restricted window. Manoogian and others do echo the conventional wisdom that consistency is best: feed your body when it's used to being fed. With time, this should correspond to when you feel hungry and when your organs are primed to deal with the insult. Fueling the body irregularly, on the other hand, can exacerbate circadian misalignment.

Consistency is an area in which I could use improvement. My unstructured freelance lifestyle has both its perks and its perils. I've tended to eat sporadically across the day, and rarely in direct response to hunger pangs. But I can remember a time when my mealtimes were more regular. One of my clearest memories of my childhood home is the analog clock that hung behind the kitchen table. On its face was the Swedish Chef from *The Muppets* making soup—with a chicken in the pot, as opposed to zooplankton. The hands of the clock were a pair of wooden spoons. And when the shorter spoon pointed toward the six, food was usually on the table.

MY MAIN TAKEAWAY: A SHORT, CONSISTENT WINDOW OF EATING, skewed early, is best. My other takeaway: you are not destined to get obese or ill if you eat late at night. My dad has enjoyed a bedtime bowl of cereal or ice cream for as long as I can remember. He's now in his eighties and incredibly fit and healthy. For what it's worth, he also eats his four meals at nearly the same time every day. Maybe the Swedish Chef deserves some credit?

Scientists still haven't elucidated all the ins and outs of time-restricted eating. Many of the most exciting results come from animals. Even some of the findings in humans conflict. Our behavior is far less predictable and more difficult to study than a rodent's. The disparities in results might be partly explained by another limitation: studies rarely consider an individual's chronotype or sleeping pattern. Until more recently, few had considered the role of age or sex.

This latter fact motivated Julie Pendergast, a circadian scientist at the University of Kentucky, to investigate sex differences. Her findings have added a twist to the story. Female mice are generally protected from many of the unhealthy diet-induced changes seen in males. When given a high-fat diet, for example, the females do not start eating at odd times like the males do. They do not get obese or sick. However, this protection appears contingent on circulating estrogens. As soon as an animal's ovaries are removed and their estrogen levels drop, Pendergast's research shows that they begin eating erratically and their risks skyrocket. If these females are then either put on a time-restricted diet or given back estrogens, they are protected once again. Her team is now looking to see whether the findings hold in humans. It's known that after women go through menopause, their risk of metabolic disease jumps about threefold. When we spoke, Pendergast and her team had just begun a clinical trial to see whether time-restricted eating might improve the metabolism of postmenopausal women and stave off these unwelcome effects.

Other researchers are considering the impacts for people who don't sleep at "normal" times. Those late-night cookie bakers and deliverers are part of a large and growing population of shift workers across the world. On top of sleeping and seeing light at unnatural times, they face the added hazard of eating at inopportune times. Plus, food options for people working the night shift are usually limited, often to vending machines. Both the poor nutrition and the poor timing can pose circadian and other health troubles.

The latest research raises a tantalizing possibility: the higher rates of heart disease, diabetes, and cancer seen in shift workers may be at least partially driven by severed ties between the brain clock and peripheral clocks ticking in the liver, pancreas, and elsewhere in the body. We know that variable timings of light and food can adjust different clocks at different rates. We know that the resulting circadian dysregulation can leave people feeling discombobulated. It can lead to inflammation and impaired immune functioning, which may promote the onset of several unwanted conditions, including cardiovascular disease, cancer, dementia, and Parkinson's disease. But what if we could nudge those clocks back into alignment—or at least reduce the discrepancies?

Restricting a shift worker's meals to daytime hours could prevent some of this uncoupling. By eating only when the circadian system is primed for a meal, even while working nights, workers might minimize the blood sugar spikes and lower the risk of developing cardiovascular and metabolic disruptions. One study of firefighters working twenty-four-hour shifts found that restricting meals to ten hours during the day reduced bad cholesterol, as well as improved blood pressure and blood sugar levels for those with risk factors for heart disease.

One particularly vulnerable subset of shift workers is pregnant women. Any strategies to rescue their rhythms will improve not only their health but their chances for delivering a healthy baby. In fact, any pregnant woman facing the pervasive hazards to our clocks could similarly benefit.

From very early in development, the mother relays time-of-day cues to the fetus through the ebb and flow of body temperature and changes in circulating hormones and nutrients. The fetus may also get information from light of varying colors and intensities penetrating the womb. Photons do get in there, believe it or not. Still, the fetus counts most on the mother to sustain strong daily rhythms of her own. The mother's SCN basically serves as a conductor of the fetus's clocks. Even if the mother lacks full control over the lights she sees and the hours she sleeps, she might more easily control when she eats. While so far only shown in rodents, time-restricted eating may reduce risks of spontaneous abortion, premature delivery, and low birth weight. More generally, according to other animal studies, maintaining circadian rhythmicity during fetal development may benefit heart health, bone mass, glucose tolerance, social behavior, and mood later in life. What's good for Mom is good for baby.

After birth, the baby's parents continue feeding supplemental circadian cues. It's hard to imagine that timing could matter much for newborns who sleep, eat, and dirty diapers around the clock. Eating many small meals across the day and night is probably necessary given the developing baby's tiny tummy. Their energy storage tank is small. But even while their own circadian and sleep homeostat systems are maturing, the baby still seeks rhythm in its days. And that rhythm remains important for its healthy development. Unfortunately, we rarely make it easy for babies to catch the beat.

I was curious to learn of one widely overlooked obstacle. New mothers commonly pump breast milk and store it for later use. Milk pumped at 9:00 a.m. and 9:00 p.m., however, can be very different milk. Just as hormones circulating throughout our bodies vary by time of day, so do the hormones in breast milk. Day milk contains more alerting cortisol; night milk contains more sleep-inducing melatonin, along with increased levels of other nighttime hormones like appetite-suppressing leptin. If pumping and feeding times are mismatched, a baby will receive conflicting cues. And that may result in compromised sleep—probably for both baby and parents.

New parents face enough pressures as it is. I am hesitant to add to their worries. But the remedies here can be relatively simple and may bring well-deserved rest for exhausted parents. Labeling bottles of "day" and "night" milk could be one low-hanging solution. Avoiding the use of melatonin-suppressing digital devices or bright blue-rich lights during night feeds is another. Research suggests that parents could also reduce the number of times their baby wakes during the night by feeding them at consistent times every day. Generally, babies have been found to develop circadian patterns, including consolidated sleep through the night, sooner when provided with strong circadian cues. What is good for baby is good for Mom and Dad.

The developing science of time-restricted eating is also being applied for older kids at Cincinnati Children's Hospital Medical Center. For any hospitalized patient who cannot eat independently, the current standard of care calls for hanging a bag of liquid filled with carbohydrates, proteins, fats, vitamins, minerals, and electrolytes—essentially, everything they need to stay alive—beside the patient and slowly infusing it around the clock for weeks to months at a time. A child with leukemia who receives a bone marrow transplant at the hospital had been kept on this twenty-four-hour intravenous regimen during their recovery, which typically takes at least a month and even as long as a year. John Hogenesch, a circadian biologist at the hospital, questioned this practice. He knew the science shows that eating at the wrong circadian phase can drive the development of obesity and hypertension, among many other health problems. And he had noticed that nearly all these children develop hypertension during their recovery. Hogenesch and his team are now collecting data to back up his hunch that shrinking the window of intravenous feeding could lead to better outcomes. "Just changing this simple thing could improve their blood pressure and metabolite rhythms and, who knows, maybe even help their cells to engraft quicker," he told me.

So far, his team has found that the kids fed for only twelve hours a day met the criteria for eating orally, the clinical gate to leaving the hospital, five to seven days sooner than those fed continuously around the clock. That could add up to savings of $100,000 per patient, Hogenesch said. More important, it could also get kids home to their families and friends, and back to eating cookies, sooner.

Round-the-clock intravenous feeding remains the standard of care for more than just pediatric bone marrow transplant patients. If the benefits in the study hold up, Hogenesch suggested the strategy could be translatable far and wide. If nothing else, ceasing the practice of sending fluids and nutrients coursing through a child or an adult during the night means their sleep is less likely to be interrupted by having to pee.

AS SCIENTISTS DECODE THE MECHANISMS RUNNING OUR RHYTHMS, they are also gaining a better understanding of just *how* pervasive pollutants and the timing and quality of other things we put in our bodies can corrupt our clocks. A popular line of inquiry is the link between the ticking of our clocks and the sway of the trillions of microbes with which we share our bodies. The composition of the gut microbiome is now believed to influence not only digestion but also the healthy functioning of our heart, immune system, and brain. We know that an imbalance in this intestinal flora may make us more prone to developing problems with these systems—like irritable bowel syndrome, diabetes, autoimmune disease, and mental illness. We now recognize a link between the same chronic conditions and circadian disruption. This may not be a coincidence. Rather, a dynamic and complex interplay between our internal clocks and our indigenous microbiota could be a force behind multiple disease processes.

Populations of gut bacteria fluctuate rhythmically across the day, possibly in anticipation of variations in the availability of nutrients and in the arrival of rival microbes. When the microbiome is confronted with a high-fat diet or irregular eating times, studies in animals and humans suggest that it can collectively lose its beat. An unhealthy imbalance can result. Gut dysbiosis can then promote glucose intolerance and obesity. Scientists even found that the health issues could be shared between mice via fecal transplantation. The good news: we are not powerless to the whims of the microbes feeding and fasting inside us. We can influence the diversity of our microbiome—welcoming more good tenants and evicting bad ones—by our exposures to light and dark, and by what we eat and when.

A healthy diet and consistent mealtimes, ideally within a reasonably restricted daytime window, could benefit the whole symbiotic community of human and microbe and potentially stave off disease. Beneficial plant compounds, such as antioxidant polyphenols, are now believed to exert a powerful influence on our gut microbiome and, in turn, our circadian clocks. And time-restricted eating appears to tip the balance of the microbial ecosystem toward more healthful bugs rather than species linked to obesity and other diseases. By fortifying a healthy microbiome that supports healthy circadian rhythms, our commensal occupants might pay rent by collaborating with our own cells to detoxify, metabolize, and otherwise mediate the physiological insults to our bodies—from particulate matter to a poorly timed pork tenderloin.

Given the apparent rhythmic influence of our microbiome, it's no wonder that antibiotics may cause our clocks to slip off beat, too. We can be exposed directly through antibiotic treatment for an infection or indirectly through the environment due to widespread use of antibiotics in humans and livestock. Antibiotics kill not only unwanted bacteria but also the good bugs. The change in the composition of the microbiome can confuse clocks.

It's complicated, with effects going in multiple directions; circadian disruption could upset the microbiome, and microbiome disruption could upset circadian rhythms. In addition to more conservative use of antibiotics, keeping a stable routine—including healthy meals at healthy times—could help the microbiome bounce back from damage wrought by antibiotics or other toxic exposures.

All of this is easier said than done in modern society, which is filled with further forces that can cause the hands of the clocks inside and outside our bodies to pass in the night. After crossing nine time zones on my return flight from Europe, I was reminded of the repercussions—the gut troubles, the mental fog, the mixed messages about when to go to bed and when to eat. And the consequences didn't stop there, unfortunately.

8

Mismatched Hours

My flight home to Seattle touched down in the late afternoon Pacific daylight saving time. Returning after three weeks in Europe, my body clocks remained anchored in Copenhagen and set at closer to midnight. I still managed to stay awake until a reasonable local bedtime that night—an easier feat than falling asleep ahead of my biological schedule when I arrived in Europe—and got what I thought was decent sleep. My eyes only momentarily sprang open at 1:30 a.m., about when the Danish were rousing. After getting up later that morning, Seattle time, I drove to the grocery store in my beloved first car, a blue Honda Civic named Boom. I stocked him full of food. Then I totaled him.

The unfortunate aptness of Boom going boom is not lost on me. The accident happened at one of the infamous blind left turns I had, perhaps with a fair bit of luck, safely navigated hundreds of times before. But was only bad luck to blame that late October morning? Probably not. Although calling it bad timing is likely fair on a couple of accounts.

Jet lag is a merciless clock wrecker. It torments us by uncoupling our biological rhythms from the sun's twenty-four-hour cycles. As the light-dark cycle shifts, our fine-tuned physiology drifts. The classic mismatch between the internal and external day can trigger hunger pangs at midnight and heavy eyes at noon. It can dampen mood, focus, and reaction time. As I learned after my crash, jet lag is known to play a role in traffic accidents, a leading cause of death for international travelers. I'm grateful to report that neither myself nor the other driver was seriously hurt in the collision. I can't say the same for the flying chicken, pumpkin, and pasta salad.

Jet lag's jarring effects linger for about a day for each time zone crossed, with assorted clocks taking more or less time to adjust. As Michael Young, the Rockefeller University circadian scientist, put it, "When you travel halfway around the world, you scatter your clocks. The day you arrive, you feel really rotten because your liver is still somewhere over the Atlantic." The human body is eventually able to reset and recover with the help of local light-dark cycles and other clock-winding zeitgebers. Evolution bestowed us with a relatively malleable circadian system, presumably to help us adapt to changing day lengths across the seasons. It's just that natural changes in the timing of daylight and darkness aren't as drastic and sudden as our man-made ones. During even the most rapidly expanding or contracting day lengths in Fairbanks, Alaska, the maximum day-to-day advance or delay in sunrise is just four minutes. Even our ancestors' journeys across oceans would have been slow enough for the circadian system to adapt—albeit rough enough to trigger miserable bouts of seasickness. With just one day of travel, however, my sunrise shifted by about nine hours.

Unless you're a flight attendant, corporate elite, or college athlete, jet lag is hopefully only an occasional nuisance. That's a relief—until you realize you don't need to get on a jet plane to throw off your clocks. Other contemporary threats are making similar jumbles of our circadian rhythms every day without our going anywhere.

FOR MOST OF HUMAN HISTORY, PEOPLE ESTIMATED THE LOCAL TIME and date from celestial bodies—the progress of the sun across the sky, the positions of the stars, the phase of the moon. The invention of the sundial in ancient Egypt sometime before 1500 BCE likely marked the first time-tracking on par with how we know it today. Sticks placed in the ground indicated the time by the length and direction of their shadows. For the following millennia, the period of daylight—from sunrise to sunset—served as the basic unit of time. The Egyptians broke that period into twelve roughly equal parts, the precursor to what we call hours. Back then, those twelve chunks varied with the length of daylight, which varied with the seasons. They also varied from place to place. Yet they all followed the sun.

Diverse cultures enlisted other tools to approximate the incremental passage of time indoors, without the need for light from the sun, moon, or stars. People burned oil and candles. They sent water dripping through holes in the bottoms of bowls and sand trickling down hourglasses. Still, these measures didn't always line up with their tracking of day and night from celestial bodies, because the durations of light and dark fluctuated greatly between summer and winter. A Greek engineer did design a complex water clock to accommodate the day-length changes. But the ultimate remedy only came with the first mechanical clocks, which began tolling hours atop towers in the thirteenth and fourteenth centuries. Finally, time's passage could be sliced into equal and discrete units and precisely counted across the day, distinct from the movement of the sun.

A separate social clock was born. It ushered in a new era of regularity, punctuality, and independence from natural cycles. It would go on to govern people's day-to-day schedules—when to arrive at work, when to attend a church service, when to watch a joust or jester or, eventually, Judge Judy—even if it didn't match the time suggested by the sun or by the body. Catholic monks built these earliest mechanical clocks, which chimed when it was time to pray. In fact, "clock" itself originates from the Latin word *clocca*, meaning "bell." And it became common to add the phrase *of the clock* when referring to time on one of these devices—hence, the *o'clock* we use today.

Over the next centuries, clock technology got increasingly good at keeping time. Yet until about a century and a half ago, we had no time zones, let alone the practice of changing our clocks twice a year. The social clock remained locally aligned with the sun. Noon still meant the sun was due north or due south—if you were in the Southern or Northern Hemisphere, respectively—and at its peak point above the horizon. Midnight still fell halfway between dusk and dawn. In most cities and towns, public clocks and bells marked the hours, and the well-to-do carried pocket watches that they synced with their town's time. If there had been a line of clocks stretched across the United States, as *The New York Times* described on November 18, 1883, "from the extreme eastern point of Maine to the extreme western point of the Pacific coast, and had each clock sounded an alarm at the hour of noon, local time, there would have been a continuous ringing from the east to the west lasting for 3 ¼ hours."

On that November day in the late nineteenth century, with chilly temperatures due in part to lingering sun-blocking ash from the eruption of Krakatau in what is now Indonesia, the *Times* was prepping its readers for another historic moment: the loosening of the tethers between the sun and social clocks. Inconsistent clock times between cities were causing chaos for the telegraph and domestic and international travel. Confusion, as well as collisions and missed connections, ran especially rampant on the train tracks. So the railroads enacted a plan to create a national system of standard time.

Most US cities and towns agreed to recalibrate their public clocks and bells that day. In New York City, the bell at St. Paul's Chapel rang twelve times for noon—twice. The first tolling came at the sun's zenith over Manhattan; the second sounded about four minutes later, marking the new noon. The next morning, on November 19, 1883, the *Times* reported on the day time stood still: "Curious people, some of whom could not exactly understand how the time could be changed without some serious results, crowded the sidewalk in front of jewelry stores and watch-repairing establishments to see the great transformation. There was a universal expression of disgust when it was discovered that all that was necessary to effect the change was to stop the clock for four minutes and then start it again." (More precisely, clocks in New York City stopped for 3 minutes and 58.38 seconds.)

From that time forward, city clocks from Maine to California—if properly set—would strike simultaneously, albeit for different hours of the day based on their designated zone. The prescribed local time of the social clock no longer aligned with the sun. In Boston, the discrepancy was about sixteen minutes. The gap in Augusta, Georgia, was thirty-two minutes. A few years later, Augusta rejected the imposed standard time and pushed the hands of the city clock forward again thirty-two minutes. As a local newspaper reported, "Augusta was again placed abreast of the sun."

More than three decades passed before US time zones were officially set with the Standard Time Act of 1918. Meanwhile, Britain was already decades into unofficially following its own standard time, until its official adoption in 1880. British time relied on an increasingly influential timepiece in Greenwich, just east of London.

A RED BALL SITS ATOP THE ROYAL OBSERVATORY, WHICH SITS ATOP A hill in Greenwich Park overlooking the river Thames and the London Docklands, a historic British trading hub. Since 1833, the observatory has hoisted and dropped the ball at precise times, a tradition later emulated by Times Square's high-LUTS New Year's Eve spectacle. Astronomers originally devised the ball drop for mariners, who could look up from the river before setting sail and accurately set their timekeepers. People could also climb the hill to check the time at any time of day.

Mounted on the outside of the brick wall surrounding the observatory is the Shepherd Gate Clock. A nearby plaque notes that this "slave dial" once received electrical pulses from a "master clock" inside the observatory. There, a state-of-the-art telescope tracked time based on the movement of stars. By 1853, the observatory's master clock was transmitting time signals to a vast network of peripheral clocks across the observatory, to the London Bridge, and along railways throughout England.

Impressively, humans created these time-telling cues and interconnected systems of timekeepers long before realizing the parallels in our own physiology. Some people even took on direct roles themselves. Ruth Belville, widely known as the Greenwich Time Lady, was among these clock runners who disseminated the master clock's time. Every week, Ruth carried an eighteenth-century pocket watch named Arnold in her handbag to the observatory. She would check Arnold's accuracy against the master clock and then travel around London to sell time to her subscribers—a tradition her father, a senior astronomer at the observatory, began in the 1830s. Business boomed for the Belvilles during and after the Industrial Revolution. Time was growing increasingly of the essence. Mechanical clocks had become household necessities and commonly adorned building exteriors. Yet most timepieces of the day weren't all that accurate, many gaining or losing several minutes a day. They needed regular tuning.

Ruth's competition intensified in the mid-1930s, with the introduction of the speaking clock. The British could now dial the letters T-I-M on a telephone in their home, or in one of those proliferate red boxes, and hear the winner of a nationwide "golden voice" contest state the precise standardized time. At least one such service still exists globally, despite the greatly improved accuracy of today's digital timekeepers. Call the US Naval Observatory at 202-762-1401 and you, too, can hear the original 1978 recordings made by actor Fred Covington: "At the tone, eastern daylight time, fifteen hours, seven minutes, thirty seconds"—*beep*. Ruth maintained a solid client base for a few years into the age of T-I-M before finally retiring Arnold in 1940. She died three years later from gas poisoning after, of all things, dimming a lamp at night to conserve fuel.

A lot changed during that pocket watch's reign. In Arnold's early days, precision timekeeping was still locally standardized. Within a few decades, it became national. Meanwhile, life in the wake of the Industrial Revolution was rapidly turning global. World leaders had begun to recognize the need to establish an international baseline for timekeeping that could support worldwide communication and travel. The local time for any place on the Earth depends upon its longitude, an imaginary line running between the North and South Poles and passing through that location. And while perpendicular latitude lines could be specified based on their distance from the equator, there was no comparably obvious "middle" for longitude. The world needed a zero to count from.

In 1884, leaders met at the International Meridian Conference in Washington, DC, and agreed that the crosshairs in the eyepiece of the Greenwich telescope would define that much-needed longitude zero. This prime meridian, as they called it, would divide the world into two equal halves. You can still stand astride the stripe in the Royal Observatory courtyard, with one foot in the east half of the planet and the other in the west half, as the line was originally defined. A new International Reference Meridian was designated in the 1980s about one hundred meters to the east of the historic prime meridian. For the last 140 years, the longitude of every place on the planet has been measured by its distance east or west from the historic or revised meridian running through Greenwich. The 1884 conference also adopted Greenwich mean time as the international basis for time zones. The result was an additional twenty-three longitudinal lines, evenly spaced relative to Greenwich. You can imagine the Earth cut into twenty-four equally sized apple slices, each representing one hour as well as 15 degrees longitude, or one twenty-fourth of the 360 degrees the Earth rotates daily. Theoretically, this meant every social clock on the planet would be set at most thirty minutes ahead or behind the local sun clock. That's not quite how it played out.

Countries were soon observing time zones relative to the new zero longitude. But many of the lines that leaders drew and redrew over the years were arbitrary, or political, and generally lacked consideration for biology. Today, more than three dozen time zones span the globe, in hodgepodge fashion. Some lines are leftovers of World War II alliances. In a gesture of solidarity with the Nazis, for example, Spain pushed its clocks ahead in 1940 to match Germany. Every summer since, the sun rises as much as two hours later in Seville than in Berlin despite their social clocks reading the same time. This clock mismatch could explain why Spaniards famously eat dinner so late; the tradition might be more biological than cultural. It could also explain why the average Spaniard sleeps nearly one hour less than the average European.

The extremes are even greater in China, which stretches across about 3,250 miles and covers more than 60 degrees of longitude. The sun can rise more than three hours earlier in the east than in the west. Yet China uses only one official time zone, set near the country's eastern edge in Beijing. Should you ever feel the need to channel your inner Marty McFly, you can cross the border from western China into Afghanistan and travel back in time by three and a half hours.

ON THE EIGHTH MORNING AFTER MY JOURNEY BACK IN TIME FROM Copenhagen to Seattle—covering a far greater distance—bright beams of autumn sunlight breached my curtains and stirred me awake. My iPhone read a couple of minutes after 7:00 a.m. I had nearly recovered from jet lag at that point and had also nearly recovered my schedule of waking up around 7:45 a.m. The day before, the sun had politely waited until 8:00 a.m. to rise. Why was it now interrupting my slumber so early? Then it dawned on me that the clock on my phone had dialed back an hour while I slept. I had actually enjoyed bonus minutes of sleep that morning. And now, for the next several months, my internal clocks would enjoy a tighter tether to the sun.

After the widespread adoption of time zones, the discrepancy between the social clock and sun clock stretched further as countries began playing with seasonal time changes: springing forward to daylight saving time and falling back to standard time every year. The modern concept of daylight saving time is widely credited to George Hudson. In 1895, the entomologist from New Zealand proposed pushing time forward by two hours so he could have more daylight after work to hunt bugs. However, the origins loosely date back more than a century earlier to a 1784 satirical essay penned by Benjamin Franklin while he lived in France.

In the *Journal de Paris*, Franklin reported awakening at 6:00 a.m. to find rays of sun pouring into his room: "Your readers, who with me have never seen any signs of sun shine before noon, and seldom regard the astronomical part of the almanac, will be as much astonished as I was, when they hear of his rising so early; and especially when I assure them that he gives light as soon as he rises. I am convinced of this. . . . I saw it with my own eyes." Franklin went on to suggest that church bells, even cannons, be used to wake Parisians and force them to start their days earlier. Using the free and "pure light of the sun," he noted, would be far better than using the "smoky, unwholesome, and enormously expensive light of candles" at night. Franklin also proposed another window tax of sorts, this one "on every window that is provided with shutters to keep out the light of the sun."

In July 1908, the residents of Port Arthur (now Thunder Bay), Ontario, became the first to implement daylight saving time. But the idea didn't catch on until Germany and Austria turned their clocks ahead an hour during World War I. By taking away one hour of morning light and adding it to the evening, they aimed to divert the coal used for electric lighting to the war effort. Other countries mirrored their model. In 1918, as part of the Standard Time Act, the US instituted its first daylight saving law to limit electricity use at night. It proved unpopular and, after the war, Congress voted to restore standard time.

The whole thing was tried again by many countries during World War II. Britain observed double daylight saving time—turning clocks two hours ahead of Greenwich mean time to save on fuel and allow workers to get home before the nightly blackouts. The US reinstated a one-hour advance of the social clock to so-called war time. This attempt, too, didn't go over well. As *Time* magazine reported in January 1942, "Farmers snarled, radiomen howled, railwaymen whistled, but Congress decided to give the U.S. daylight-saving time—the year round, for the duration." An article from the magazine printed a year earlier even referenced the repercussions for "the psyche of the U.S. cow." Objections to the daylight saving time proposal, noted the magazine, included "the assertion that the cow is no dull creature of sodden disposition . . . such as an hour's change in the milking time, might make a cow tense, thereby impairing the flow of milk necessary for national defense."

As soon as the war ended, officials repealed the law and individual states could go back to standard time. Some regions, however, chose to remain on daylight saving time in spring and summer months. The clock-changing schemes varied within and between states, leading to widespread confusion and eventually compelling Congress to pass the Uniform Time Act in 1966. The law established daylight saving time from the last Sunday in April to the last Sunday in October. While states could opt out and stay on standard time year round, as most of Arizona and Hawaii would do, now most of the country was changing clocks twice a year.

The US government was not yet done playing with the clocks on Americans' walls. In 1974, it again imposed permanent year-round daylight saving time as a purported energy-saving measure during the oil crisis. Perhaps because a generation had passed since the strategy's last failure, the move initially garnered popular support. But Americans' enthusiasm waned. *The New York Times* described a sample of the ensuing consequences, from a garbage collector unable to see trash cans to a suburban family forced to start the day in the "pitch black" after Dad blew a fuse making toast. The country, according to the writer, had entered the "Second Dark Age." Even more concerning, kids were getting killed on their way to school in the dark. Officials lifted permanent daylight saving time ten months later, and most clocks in the US once again began flip-flopping twice a year. The expected energy savings never manifested.

Since World War I, more than 140 countries—primarily those a decent distance from the equator, where seasonal day-length changes are more pronounced—have tried annual switches to and from daylight saving time; more than half abandoned it. Many restored standard time. However, Iceland and a few other countries stopped their clocks forward an hour and never looked back. The US is inching closer to the same as I write this book.

Legislation over the years has repeatedly lengthened the period that Americans' clocks are set to daylight saving time. A coalition of several industries—from amusement parks to petroleum corporations to sporting goods manufacturers—lobbied for the extensions. US senators from Idaho were among supporters who voted for, and won, a more than three-week-earlier April start date for the switch from standard time in 1987. Hardee's and McDonald's reportedly sold more french fries during daylight saving time, and Idaho makes a lot of potatoes. The golf industry, too, promoted the expansion of daylight saving time—anticipating more after-work golfers with the lighter later hours. Candy companies were especially pleased with the Energy Policy Act of 2005, which pushed the date for the change back to after Halloween. The act also advanced the start date to early March. Since that statute went into effect in 2007, most clocks in the US display daylight saving time for about eight months of the year.

TODAY, THREE CLOCKS RULE OUR LIVES: THE SOCIAL CLOCK, THE sun clock, and the body clock. Standard time, compared with daylight saving time, keeps the social and body clocks more closely hitched to the sun clock. Springing forward, on the other hand, will push the hour hand of the social clock unnaturally forward. In trying to keep up with the social clock after the change, your body clock now falls out of step with the sun clock.

The impacts may be both immediate and long term. The days and weeks after springing forward bring slightly greater risks of accidents, suicides, strokes, pregnancy loss among IVF patients, and other unwanted events. A US study found that the rate of heart attacks jumped 24 percent the Monday after the abrupt spring switch. Scientists even measured reductions in altruistic giving and found judges more liable to dole out longer sentences after Americans set clocks forward an hour. And then there is simply the confusion and annoyance many of us face twice a year: Which way do I turn the clocks? When? *Why?*

Yet maybe more concerning are the subtle, chronic effects of daylight saving time. Our bodies never fully adapt to the spring change. Leaving our clocks ahead an hour means that, every day, we are more apt to miss a valuable hour of clock-entraining morning light and gain an hour of clock-derailing late-evening light. Researchers have devised an ingenious method to estimate the impact of reallocating that hour of light: compare the health and productivity of people on opposite edges of time zones. One team of economists estimated that the circadian misalignment that comes from living on the western side, where sunrise occurs at a later clock-on-the-wall time, results in an average 3 percent lower wage compared with the eastern, earlier sunrise side. Rates of suicide and car accidents are higher on the western side, too. The risk of cancer even climbs slightly from east to west within a time zone. The reverse pattern looks to be true for longevity: fewer years farther west. Most immediate of the repercussions, concentrating natural light later in the day was found to rob the average person of about nineteen minutes of sleep each night.

Springing forward an hour mimics a move from the east to west edge of a time zone. The resulting discord between the social, sun, and body clocks exacerbates social jet lag for everyone—and worsens the plight of people already living on the wrong side of a time zone.

The sudden change can botch up our internal calendars, too. That one hour translates to weeks of a gradual seasonal shift, experienced in the shrinking or expanding gap in time between sunrise and the start of school or a job. Sunrise naturally creeps earlier in the spring before the fast-forward to daylight saving time. Then, slowly, the timing of sunrise rewinds again relative to the social clock. The reverse happens in the fall. In Seattle, the sun waits to show its face until about 8:00 a.m. daylight saving time in early November—as it did during that first week after my return from Europe. Then there's that fall-back morning when we gain an hour of sleep. Many of us Seattleites return to rising with or after the sun, albeit temporarily. Within a few weeks, the shrinking day lengths push the sunrise again to nearly 8:00 a.m., now standard time. As I sit at my desk typing this at 7:50 a.m. in late December, it would be almost pitch black out my window if not for the LED streetlights. And if year-round daylight saving time were to be instituted again, the darkness would drag on even longer—until almost 9:00 a.m. for several weeks of the year.

These dark mornings disproportionately affect underserved groups, who often have less control over when they start their days. Middle and high school students are hit hard, too. Despite the data and history lessons, and the pleas from the sleep and circadian research community, a campaign is underway as I write to once again permanently turn clocks forward an hour in the US, Canada, and the European Union. As one expert said, "People have short memories."

THERE IS ONE FUNDAMENTAL REASON WHY OUR SEASONAL TIME changes matter: our daily lives are intimately hitched to the social clock. The advent of the social clock, facilitated by the arrival of mechanical clocks, introduced possibilities for strict schedules. Mismatches between the social, sun, and body clocks quickly followed. Today, the artificially constructed times that society expects us to show up for school or work rarely fit the internal clocks of students and employees. The suffering starts early.

For Ellen Jatul, it began between the seventh and eighth grades. She had spent her youth falling asleep and waking up reasonably early. Then came the change. "I started to have trouble going to bed early. It was like my body just wouldn't shut down," she told me. Yet now she had to wake up early; her Seattle middle school started at 7:50 a.m.

Ellen would often wear her school clothes to bed so she could roll out, grab a bagel, and leave her house by 7:15 a.m.—the biological equivalent of about 5:15 a.m., or earlier, for an adult—all after needing multiple alarms to wake up. "I would almost have to run to school," she said. A good portion of the year, her mile trek to school would be in the dark: "Half the time I don't even remember walking, 'cause I was so tired." Then, on the weekends, she could sleep in until noon. Needless to say, Ellen's social jet lag was high.

Tradition largely dictates the biologically backward school times—with morning bells ringing later in elementary schools than in middle and high schools. I remember starting around 9:00 a.m. at Emily Dickinson Elementary and 7:30 a.m. at Redmond High School. Little has changed over the intervening decades. More than 40 percent of high schools in the US currently start before 8:00 a.m.

Cindy Jatul watched the students in her high school biology classroom along with her own kids—Ellen included—struggle to conform. Especially during a first period that started before 8:00 a.m., she told me, they were simply not functional. "We need to recognize the importance of biology and try to work with it, instead of trying to just ignore it," said Jatul, now retired from teaching and a previous career as a nurse practitioner. "Time after time, biology comes back and says, 'No.'"

During middle and high school years, sleep pressure begins to build up more slowly through the day. The circadian system also creeps later. These changes conspire with the extended evening daylight after springing forward, and with technology and social factors, to keep older kids awake well into the night. People have recognized the phenomenon for at least a hundred years. Before social media, many faulted late hours on video games or television. Before that, dancing bore much of the blame. So, is the issue behavior or biology? "I think it's a little bit of both," said Mary Carskadon, a professor of psychiatry and human behavior at

Brown University and a renowned child and adolescent sleep expert. One telling clue: teenagers' predilection to stay up late isn't exclusive to modern or urban societies, nor is it exclusive to humans. But it can be exaggerated. And let's not forget about the additional impacts from late-night blue light and exercise. Bottom line, when a teen is expected to be at school before 8:00 a.m., there's almost no avoiding sleep deprivation. "Kids may be sitting in schools, but their heads are still asleep on pillows at home," Carskadon told me.

The American Academy of Pediatrics recommends between eight and a half and nine and a half hours of sleep for middle and high school students. Fewer than one in four US high school students sleep eight hours per night. The average is around seven hours. Any parent can attest to the resulting morning struggles. Ellen admitted that both she and her sister would spend the mornings grumpy. But the problems go deeper. Sleep deprivation carries numerous costs for teens, including lower grades and higher rates of car accidents, risky behaviors, depression, and anxiety. Cycles of REM sleep primarily occur in the last third of the night. By cutting a night of sleep short by an hour or two or more, a teenager loses this vital time for the brain to solidify learning into memories and process emotions. REM sleep may play a critical role in strengthening critical thinking and problem-solving skills, too. "Their brains are craving REM sleep," Carskadon told me. Because the body releases growth hormone during sleep, sleep deprivation could even stunt a kid's growth over time.

Once again, these impacts aren't equally distributed. Latitude can make a difference. The public high school graduation rate in Alaska is just 79 percent, for example, well below the US average of 87 percent. Clay Triplehorn, the sleep doctor I met in Fairbanks, places partial blame on early start times combined with extreme day-length changes. For many months of the year, students "get no morning light to set their circadian timing," he said. Socioeconomic factors also play a role. Teachers shared with me how lower-income students commonly care for their siblings in the mornings and rely on public transit to get to school—forcing them to set alarms even earlier. Then, of course, there are the differences in our biology.

Starting as early as elementary school, later chronotypes have been found to get poorer grades compared with earlier chronotypes, especially in the sciences. A Dutch study found the effect comparable to absenteeism. The disadvantages continue as kids grow into adults and enter the working world. Night owls may be more than twice as likely to underperform at work, and night owl men may be three times more likely to go on disability compared with morning larks. On average, owls earn lower incomes than larks, which researchers peg in part to associated lifestyle choices more common for owls, such as drinking alcohol, smoking, eating unhealthy foods, and exercising too little. These choices are arguably influenced by a forced adherence to the early social clock, and resulting accumulations of social jet lag. Benjamin Smarr, the data scientist from San Diego, pointed out that people predisposed to being night owls end up "chronically disadvantaged across their entire education." This ultimately affects their long-term employment and well-being. "If there were a weight restriction or a height restriction on how you get education, people would complain, right?" he said. "We haven't had a great way to measure [chronotype], so people just ignore it. But it's real."

A one-size-fits-all work schedule has dominated since the Industrial Revolution. Social jet lag and all its consequences soon followed, with workers of all chronotypes—and their employers—paying the price. Research shows that disrupted body clocks can lead to deficient work performance. We may get sick more often, too, either directly due to lack of sleep or simply because we sleep and eat at the wrong times for our bodies.

With the introduction of electric light, companies also soon realized they could make more money by keeping workers on the production line around the clock. Clock-in and clock-out times started stretching into late-night and early morning hours. The swing shift and the night shift were born. Today, the proportion of people employed in shift work continues to rise to meet the demands of a round-the-clock global economy. As more countries add electric lighting, experts anticipate the population of people working nontraditional hours will grow even further.

For the approximately one in five people in industrialized countries engaged in shift work today, the afflictions of social jet lag and sleep deprivation can be extreme and chronic. Truck drivers, doctors, nurses, first responders, janitors, store clerks, shipping operators, and flight crew members are just a few of the people who may work into the night, early morning, or on rotating schedules. Whether they realize it or not, these people often work, eat, and sleep against their natural clocks. The insults can accumulate, wearing down their bodies and raising risks of various maladies.

LANELLE SCHULTZ AND LISA FINELLI ARE TRUCK DRIVERS AND LONG-time friends. Schultz pulls fifty-three-foot trailers across large swaths of the country. Finelli transports whatever UPS asks her to around the Seattle area. Both do much of their driving through the night.

They call themselves "vampire drivers." They refer to anybody working daytime hours as "daywalkers." However, the friends' schedules routinely diverge. While Finelli tries to maintain her vampire hours seven days a week, supported by a husband who also works nights, Schultz told me that her schedule drifts and often ends up flipped by the end of the week.

Schultz's experience is common among transportation workers, according to Ryan Olson, an occupational health researcher at the Oregon Health & Science University. In addition to recommending cabin enhancements for drivers, such as suspension seats and therapeutic mattresses, he is encouraging drivers and employers to improve the consistency of workers' hours. "We have changed working conditions to a degree but not schedules," Olson told me. "Schedules are the elephant in the room." And schedules are frequently outside a worker's control. Meanwhile, employers face the reality of slim profit margins. Staying ahead of the competition means keeping trucks on the road at all hours. "I don't know if human society asks the question enough," Olson told me. "Do we really need to run operations twenty-four hours a day? We're not prioritizing workers' well-being. We're prioritizing the convenience of customers." I can't help but think about the printer ink I ordered yesterday afternoon on Amazon and found on my porch this morning.

Naturally, other issues crop up in the real world of shift work. Finelli and Schultz complained that most hotels cater to daywalkers by offering only afternoon check-in times. "I don't need a hotel at night," said Schultz. "And I don't want to pay for two days if I only need one."

The fundamental issue with shift work is much like with the jet lag I recently endured: our circadian system was not designed to accommodate rapid changes in daily cycles. These workers are active, sleeping, and eating at the wrong circadian phases in their body. Clocks scatter and organs scramble to deal with the fallout. However, unlike with international travel, shift workers are rarely able to fully reset their internal clocks in harmony with the sleep and wake schedules required by their jobs.

Constantly disobeying inner clocks carries acute and chronic consequences. It can make shift workers more prone to mistakes and accidents on the job, and more at risk of being in a car crash on the way home. Risks are particularly high when the commute corresponds with the circadian dip in alertness—that dangerous period when core body temperature bottoms out. The hazards linger, too.

Shift workers are also now thought to face long-term risks of infertility, obesity, diabetes, heart disease, gastrointestinal problems, depression, anxiety, asthma, and other conditions. One study found hospitalized shift workers were more than twice as likely to test positive for COVID-19 than hospitalized people who work regular hours. Other studies hint that shift work early in life may elevate the risk of a severe stroke in middle age, and that shift work in middle age or later years could raise the risk of cognitive impairment. "It's almost to the point in my mind that shift workers should receive hazard pay for what they do," one scientist told me.

The World Health Organization has gone as far as to declare night shift work a probable carcinogen, but with ample caveats. Some studies do point to slightly elevated risks of breast, prostate, liver, lung, and colon cancers. But other research fails to find an effect. Aziz Sancar, a biochemist at the University of North Carolina and winner of a Nobel Prize in Chemistry for his work on the mechanisms of DNA repair, remains unconvinced of a link. He told me that proving circadian disruption from shift work can promote cancer would take a long-term study similar to the one done to confirm that smoking causes lung cancer. As he and the University of Washington's Russell Van Gelder point out in a 2021 review paper, the fact that gene expression, cell division, and DNA repair are all modulated by the clock and all involved in cancer makes it plausible that fractured clocks could predispose people to cancer. Still, the conflicting results simply don't support a causal link. Sancar and Van Gelder do, however, report strong evidence that "desynchrony between the internal circadian and external geophysical clock" causes "jet lag, sleep phase syndromes, and metabolic syndrome."

Shift workers are not the only ones who pay the price. The consequences could also extend to workers' kids, and perhaps their kids. Pregnant women working the night shift, and experiencing circadian disruption, may pass on long-term health problems to their offspring. One study of pregnant mice found that chronic circadian disruption led to mood disorders in their pups and grandpups and possibly subsequent generations. And Mom's clocks probably aren't solely to blame. Dad, too, could transfer troubles if his clocks are off-kilter at the time of conception, according to another study.

People alongside shift workers on the road or in a plane, or who are under a shift worker's care, may also be at risk. You probably don't want to fly with a clock-compromised pilot or go under the knife of a jet-lagged surgeon. Rates of medical errors, a leading cause of death in the US, are higher for night shifters compared with day shifters. One study of nurses estimated an error rate difference of 44 percent.

Exhaustion from long hours is another major driver of medical mistakes. Paria Wilson, an emergency medicine physician at Cincinnati Children's, has been on a mission to remedy the burnout culture of medicine. The prevalence of burnout for US physicians has long hovered near 50 percent and jumped to nearly 63 percent in 2021 during the pandemic. In case the consequences aren't obvious enough, she is collecting wearable data from residents to prove that "it is not okay to work one hundred hours a week and not sleep," she told me. The well-being of not only residents and physicians but also the patients under their care is at stake. The data are now helping to optimize rotations, matching residents with schedules based on chronotypes. But it's been hard to achieve change with baby boomers making the policies, lamented Wilson, who graduated from medical school in 2010. "It's just such a generational shift," she said. "We're not going to torture ourselves just because you guys did all of those things." A similar suck-it-up attitude toward schedules is widespread among maritime pilots and in the military.

MARITIME PILOTS BELONG TO THE UNLUCKY CLUB OF SLEEP DRIFTers like Schultz. These skilled mariners navigate ships in and out of their local waterways. They're like aircraft pilots whose specialty is the takeoff and landing. The pilots who take over vessels as they pass through the waterways of Puget Sound in Washington State have historically worked on a strict rotation twenty-four hours a day, seven days a week. A board in the pilot station in Port Angeles, Washington, maintains this dispatch list. As boats come in, the pilot on the top of the list goes out. Then their name goes to the bottom of the list. "Sometimes you'll end up with a lot of time between assignments, and sometimes you're what we call 'turn and burn,'" explained Captain Ivan Carlson Jr., president of

Puget Sound Pilots. He told me that his organization explored allocating pilots to daytime and nighttime crews but determined it would not be feasible. Because ships are coming in at all hours, day and night shifts would "really bleed together." So the pilots continue to tough it out, just as Kevin Brockman did while serving in the US Navy.

Brockman said he rarely got adequate sleep, let alone sleep when his body desired it. He first worked rotating shifts during his land-based training. The pattern was grueling: It started with a 7:00 a.m. to 7:00 p.m. shift for a week. Then, after two or three days off, he would work 11:00 a.m. to 11:00 p.m. for a week. After another thirty-six hours off, his schedule changed again: 7:00 p.m. to 7:00 a.m. That long week was followed by another three days off, and finally a ten-hour shift that spanned from 7:00 a.m. to 5:00 p.m. A four-day break bridged the start of another cycle. "That was fun," he told me. I asked him how he managed. Did the military offer advice or training to mitigate the jet lag and sleep deprivation? "No, we were pretty much just told to figure it out," he said. "It was very much, 'This is your schedule. This is what you're going to do. Yes, it sucks. We all know it sucks. And you're all going to do it. Do you have any questions?'"

After about two years of training, Brockman got on a submarine. Gratefully, his underwater shifts remained generally steady while the vessel was underway. Each of three teams "stood watch" for the same eight-hour chunk of each twenty-four-hour day. But those workdays would continue for hours afterward with other tasks. Brockman said he averaged three to five hours of sleep a night for the first year or two. It didn't help that he slept on a short, hard bed. "At one point, my head or my feet or both were touching a wall," he told me. To compensate for that lack of sleep, Brockman said he drank "lots and lots of coffee," the precise quantity depending on his shift. He could limit himself to one or two sixteen-ounce thermoses with a day or swing shift. On the night shift, especially during the first days that the submarine was underway, he recalled drinking coffee through the entire eight hours. "I was drinking it like it was water," he said.

On top of the long hours and crazy shifts, Brockman and the other submariners also dealt with a near-complete lack of light-dark contrast. In most rooms, fluorescent lights stayed on 24-7. The only daylight Brockman saw for weeks at a time was through a periscope—and that was via a video screen. The sole attempted compensation on board was "one of those cheap, tropical beach calendars," said Brockman. To boost morale, his boss had bought one, ripped out the pages, and taped them to the seafoam-green-painted walls of the submarine.

Brockman's circadian and sleep homeostat systems were taking hits from all sides. Yet Brockman might still have considered himself lucky. Before 2014, the typical submariner in the US Navy lived an eighteen-hour day: serving a six-hour watch and then completing other duties and sleeping during the following twelve hours, before repeating the cycle the next day—now moved six hours. Shift workers in many more professions are plagued by similarly chaotic schedules that rotate throughout the days and weeks. Even with consistent night shifts, family and social lives can drive a schedule swing on the weekends. This lack of regularity can be really detrimental to rhythms. The fact that Brockman never had the opportunity to see the sun or go home and adapt to the schedule of his family or friends was both a blessing and a curse. He was one of those rare workers who may have at least succeeded in temporarily becoming nocturnal.

A VARIETY OF CIRCUMSTANCES CAN DRAW PEOPLE TO SHIFT WORK. For the rare few, late working hours suit their biology. For more, it offers an opportunity for a higher wage without needing a pricey higher educational degree. Still others appreciate the flexibility to make ends meet while juggling obligations, such as raising kids. In many of these cases, the decision to become a shift worker isn't really a choice. And it doesn't necessarily even leave people financially comfortable. A study in the UK found that seven in ten night shift workers earned less than the median wage.

Erika Garcia works nights as an emergency room tech assistant in Grosse Pointe Farms, Michigan. When we spoke on a late afternoon in mid-January, she had only just gotten out of bed, having nearly missed all the day's natural light. She told me that she chose the night shift primarily to help cover childcare. Her husband works days. On her days off, Garcia said she tries to match her family's schedule. But she can tell that the flip-flopping takes a toll on her clocks. After her first shift back at work, she will try, unsuccessfully, to take a nap in the afternoon: "My body is like, 'What are we doing?'" She told me that she often feels cranky and irritable. She was also started on medication for heart palpitations after she began working the night shift.

Almost all of us will break our clocks to some extent at some time or another—whether we are a night owl working eight to five, waking up for an early first-period class, partying late on a Friday night, living on an extreme side of a time zone or in a misaligned time zone, flying back from overseas, or pulling late nights while writing a book about clock disruptions. The toll of our modern menaces, acting solely and in combination, is staggering. We now know that if our rhythms are out of whack, so are we. But there is hope, I promise. And it doesn't need to involve moving everyone back to the equator or reverting to life without electric lights, refrigerators, and twenty-four-hour medical care. The rest of this book will highlight approaches to minimize the harms we have been inflicting on ourselves and to maximize the benefits the new science avails to us—from loosening work and school start times and addressing shift-work hazards, to abolishing daylight saving time and correcting unhealthy time zones. We'll touch on the benefits of considering circadian rhythms in urban planning, building regulations, interior design, meal timing, medicine, and agriculture. We'll revisit simple but powerful rules that help ensure our tickers stay on a healthy beat. And we'll discover some relatively easy hacks to boost performance, productivity, and perhaps even longevity. It all starts with recreating the consistency and contrasts that our circadian rhythms crave. Our clocks are ticked off. It's time to reset.

Part III

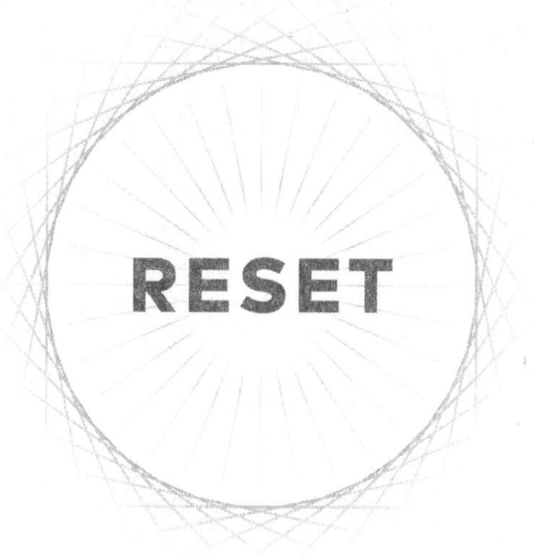

9

Goodbye, Alarm Clock

In a photograph taken in 1931, Mary Smith stands on a cobblestone street. Her right hand is on her hip, and her left hand holds a long, slender pipe in her mouth. The pipe is pointed slightly skyward—toward the bedroom window of someone she is about to "knock up."

Moments after the photo was taken, Smith presumably blew into her pipe and shot dried peas that clinked and clanked and roused the Londoner on the other side of the glass. Then she may well have reached into the pocket of her cardigan sweater, which puckers in the picture as if it holds a stockpile of ammunition, reloaded, and proceeded to knock up her next client. People actually paid her to do this—reportedly, sixpence a week—as they did her competitors, who used assorted tactics to tickle windows, including long bamboo sticks and fishing poles fitted with small pieces of wood or whalebone. It was money well spent. The choice for many working-class people at the time was to hire someone to wake them up or risk oversleeping, being late to work at the factory, and losing their job.

Before the Industrial Revolution, people could generally set the pace of their days. Working hours were flexible and varied and tended not to require rude awakenings—minus maybe an unwelcome cock-a-doodle-doo or church bell. Until the arrival of electric light, those hours also tended to align with the sun; people needed to see to do their work. But the transition from farm to factory ushered in a more regimented era, one less organized around daily and seasonal rhythms. Increasingly, workers needed "to wake according to the clock, rather than the sunrise," writes historian Arunima Datta. And this "shift in the rhythm of labour and sense of time," she suggests, opened up an "economy of waking."

While mechanical clocks were becoming more prevalent around this time, they were not yet reliable—hence, the demand for Ruth Belville, the Greenwich Time Lady. And even after an alarm function was patented in 1847 that allowed users to set a personal wake-up time—albeit only to the hour—few could afford the investment. A market grew for human alarm clocks. Many of the first "knocker ups," as they were called in England, did indeed knock on doors to wake workers. But that strategy often provided a free alarm clock for neighbors, robbing the knocker up of potential business. Either that, or it upset neighbors who would rather not be disturbed from their slumber. The tap of a pole or peas on windows—those not bricked up—solved the problem. In Italy, however, "hooters" continued to employ less gentle methods, such as blowing shrill whistles. I'm exercising so much restraint here.

Mary Smith's daughter Molly Moore would join in the business, too. While the Belville family traveled around London selling the new social time, the Smith family went about town shooting peas at windowpanes to keep people obeying the time on that new social clock, and inevitably disobeying their body clocks. Knocker ups varied their rates by the hour of awakening and by the season. Earlier hours generally meant higher prices. And greater demand during the short winter days could increase charges nearly fourfold.

Eventually, alarm clock technology advanced and prices dropped, helped along by the lifting of taxes on watches and clocks. People began to take down the "wake-up slates" that hung on homes to mark the hour of desired awakening. By the mid-1900s, the tickles and taps on windows gave way to the buzzes and beeps of personal alarm clocks, but only after several decades of resistance. One critic, writing under the pen name Anti-Buzzer in the 1870s, had referred to the new devices as "devil screamers." That still feels apt. Of course, today it seems everything has an alarm clock. Our smartphones have made the traditional stand-alone versions nearly as obsolete as knocker ups. Although there remain endless options—like Clocky, an alarm clock on wheels that leaps off your nightstand and rolls away, forcing you to chase it down as it incessantly beeps and bounds.

Multiple circadian scientists imparted their hope that we can one day render *all* alarm clocks obsolete. Impractical? Probably. But I grew convinced that they have a point. Waking up to an alarm clock rather than naturally with our internal timer is not only annoying—especially if that clock starts careening away from you—but is also not good for us. There was a time before both human and man-made alarm clocks, and there could be a time after.

I BROKE MY PERSONAL RECORD FOR LONGEST STINT WITHOUT USING an alarm clock—at least as an adult—with those ten nights in the bunker. I have since tried to avoid setting an alarm whenever possible. Given my profession, I am fortunate to have that option more often than my friends who are required to report to work most weekday mornings. While COVID-19 shook things up, many employers still enforce the traditional nine-to-five-ish routine that originated during the Industrial Revolution.

Based on the bell curve of chronotypes, the average American or European today wakes up most naturally between 8:00 a.m. and 9:00 a.m.—after going to bed around midnight. Clocking in for work at 9:00 a.m. is then obviously not ideal. Even less ideal is 8:00 a.m., a more typical average start time. Taken together, it's unsurprising that most people rely on an alarm to get to work on time.

Despite the greater number of owls and doves, strong biases remain in favor of larks. These are reinforced by popular proverbs such as "The early bird gets the worm," and "Early to bed and early to rise makes a man healthy, wealthy, and wise." The latter was endorsed by the same Benjamin Franklin who went on to write of waking everyone up earlier with booming cannons—not rattling peas. People who naturally start their days later, especially those who roll in "late" to the office, are regularly shamed and pegged as lazy. I don't think I've met anyone bothered more by this stigma than Camilla Kring.

In the early 2000s, when Kring began her PhD studies in work-life balance, scientists had recently identified the first genes that govern mammals' internal clocks. She was intrigued. Circadian rhythms have remained core to her work ever since. Kring ultimately founded the Copenhagen-based B-Society with the goal of increasing the quality of life and productivity of B-persons, her term for people with later chronotypes. As you might guess, she calls morning larks A-persons. Again, because traditional work and school hours favor larks, owls tend to suffer worse social jet lag throughout life. We now know the consequences can include far more than mental exhaustion, depressed mood, and higher consumption of caffeine and alcohol. Kring resents that society often focuses on fixing B-persons. Instead, she argues workplaces and schools should accommodate everyone from the extreme early birds to the late night owls: "Human beings should have a right to live more in sync with their biological clocks," Kring told me. She has made it her mission to create a "new time architecture."

She is preaching to the choir. I may not be an extreme night owl, but I've resented early alarms my whole life—especially during my teen years, when I probably was a true night owl. I despised getting up before 6:00 a.m. in high school. I was always frustrated by the early morning starts of 5K races. All the major tests I've taken have begun far too early for my liking. To try to adapt to society's early bird bias, I've enlisted melatonin pills, chamomile tea, new pillows and mattresses, earplugs, calming sleep apps, and alarm clocks with harsh buzzers, melodic chimes, chirping birds, and even U2's "Beautiful Day." I've also consumed an inordinate amount of caffeine. I was probably on par with Brockman's submarine coffee intake before being enlightened about caffeine's proclivity to linger in the body and inhibit sleep.

After initially working a nine-to-five career, I gratefully found one that allows me to set my own hours. Kring thinks all workers should have that flexibility. She is steering several companies—including medical giants Medtronic, Abbott, and AbbVie—toward a new work design that helps all workers better live by their inner clocks. She encourages workplaces to match schedules with chronotypes and to hold meetings midday, before early risers leave the office and after late risers arrive. Spreading out work hours carries other benefits, like reducing commute times and traffic jams. After applying Kring's concept years before the pandemic, many employees at Novo Nordisk in Denmark halved their time on the road. Postponing a meeting just thirty minutes later in the morning—say, from 9:00 a.m. to 9:30 a.m.—could often do the trick, according to Kring.

AbbVie in Norway implemented Kring's advice in 2008. Before the move, many employees reported being unhappy. Turnover was high, and the company's reputation within the industry was low. With Kring's help, leadership began allowing employees to arrange their days as they liked, so long as they delivered on their jobs. Meetings were rarely scheduled before 10:00 a.m. or after 4:00 p.m. The changes seemed to pay off for both the employees and the employer. Turnover and use of sick leave plummeted. The rate at which employees were satisfied with their work-life balance increased from 58 percent to 95 percent. Among medium-size companies, AbbVie Norway was voted the best place to work in Norway in 2014, 2017, and 2018. Marte Fjelle, the company's human resources director at the time, said that while some employers may fear that their workers would exploit such flexibility, only 1 or 2 percent attempted to do that at her workplace. "We decided to address those individuals rather than taking away the benefit from the whole organization," she told me.

With the freedom to experiment, employees also began recognizing the hours they worked their best. Fjelle discovered that her own peak time fell between 3:00 p.m. and 6:00 p.m., almost precisely when I'm in my daytime trough. During those hours, Fjelle said, she was "double as effective" as at 9:00 a.m. "You can arrange your life for the benefit of yourself and also for your employer," she said. "Two plus two was not four anymore; it was actually five."

In 2019, Fjelle moved to a new company where the owner still required traditional office hours. That was, until the pandemic hit. When the owner realized that his employees were continuing to produce while working from home and at varying hours, Fjelle told me, he too loosened his reins.

MAGNE SKRAM HEGERBERG, ANOTHER OF KRING'S CLIENTS AND general secretary for the Norwegian Association of Lawyers, learned the benefits of paying his employees for their skills and productivity rather than for their time. He switched his emphasis from the quantity to the quality of work hours. Shortly after he joined the association, Hegerberg told me, he organized a "funeral" for the clock-in machine. It had been mounted on the wall by the entrance for employees to sign in every morning and out every afternoon. He described how they tore it down from the wall, then proceeded with a "condolence protocol," complete with candles.

The starting times of the forty-plus employees at the trade union headquarters now range from 6:30 a.m. to 2:30 p.m. Tailoring work schedules to chronotypes, he said, has doubled productivity in some areas and, more broadly, improved innovation, creativity, and problem-solving. He mentioned one employee who sometimes arrives at work around 2:00 p.m., and sometimes not at all. "But I am so sure that he is delivering the best I can get," said Hegerberg.

Till Roenneberg, the chronobiologist at LMU Munich, approves the business strategy. "If I want to have a perfect workforce, then I would tell them to come when they have woken up without an alarm," he told me. Roenneberg observed significant improvements in sleep and about an hour reduction in social jet lag among factory workers when he and colleagues implemented chronotype-adjusted schedules.

With a business's value now about 90 percent intangible, companies today pay for employees' brainpower far more than for material goods. Hegerberg has also adopted one of Kring's material tools for utilizing peak brainpower: an army of plush frogs. During their power hours, an employee will place a brightly colored frog on their desk, or on the door to their office, to inform others that they do not want to be interrupted. In Kring's words, it tells others to "frog off!" A black-and-white-and-red frog from Kring came home with me from Europe. I work solo. The souvenir sits on my desk as a productivity prompt and pairs well with the late morning hours I now spend with my phone silenced.

Ultimately, Kring wants to see discrimination based on chronotype added to the human right declarations from the European Union and the United Nations: "You are born with this rhythm. It's not something you choose." And consideration of this fundamental right, she argues, should begin well before and extend to well after our working days. When I went to visit her in Denmark, Kring drove me through farms and fall foliage to a nursing home about an hour and a half southeast of Copenhagen. GuldBoSund Friplejehjem og Rehabilitering is nestled amid lush green fields, overlooking Guldborgsund, the strait between two Danish islands. Here, up to fifty residents enjoy the views, ample daylight from massive windows and skylights, and wake-up and eating schedules designed to fit their circadian rhythms—from the retired police officer who gets up at 5:30 a.m. to the woman who prefers not to be woken until 9:30 a.m. Before Kring consulted with the nursing home, staff didn't ask residents about their preferred sleep and meal schedules. Rather, workers just started at one end of the home every morning and worked their way to the other. Now, before the night shift hands off to the day shift, they make the retired officer a cup of coffee and his favorite toast with cheese. And then the day shift proceeds with all the other visits at the residents' chosen times.

I met one resident, eighty-eight-year-old Bent, in his apartment. He sat in a big black chair next to a sliding glass door to his deck. With the help of a translator, I learned that Bent had long owned a plumbing company. Because he had employees who started work at 7:00 a.m., he always had to get up early. Not any longer. Now, Bent chooses to get up at 9:00 a.m. "I'm going to bed when it fits me," he said. "Not when someone tells me to."

The staff here, too, have benefited from new schedules friendlier to their own rhythms. On average, they take fewer than two sick days a year, a significant drop from previous years. That figure includes those working the night shift.

WE CAN ONLY FATHOM A GUESS AS TO WHETHER THE EARLY WORKing hours for knocker ups, like Mary Smith, matched their chronotypes. Historical reports suggest they would typically stay up all night and start knocking as early as 3:00 a.m. to awaken workers for the first shifts. It's not far-fetched to think night owls might have been drawn to the profession, though at least one knocker up reportedly hired someone to knock him up. A popular tongue twister of the day alluded to this hiccup: "We had a knocker up, and our knocker up had a knocker up, and our knocker up's knocker up didn't knock our knocker up. So our knocker up did not knock us up, 'cos he's not up!"

As research continues to sort out the various effects of circadian disruption on shift workers, experts are also developing countermeasures. They believe that the burden can and should be significantly lessened.

Perhaps the most obvious strategy is to match extreme hours to extreme chronotypes—ideally, via self-selection, as the knocker ups presumably did. Lisa Finelli and Lanelle Schultz, the truck drivers, are both self-declared night owls who gravitated to a career that allows them to work through the night. The same motivation led Ashley, the veterinary pathologist, to seek her overnight position. Meanwhile, some employers appear to recognize the potential win-win. Southwest Airlines advertises its consideration of circadian rhythms in organizing a.m. and p.m. flight patterns. "Whether you're an early bird or a night owl," reads their pilot recruitment page, "we have something to suit you!"

The notion of a job for every chronotype is appealing, but things can get complicated quickly. "There may be a time in the future where we're able to find some people who may be more resistant to the untoward effects of shift work than others. And maybe people could assess that prior to making a career choice," said Nathaniel Watson, codirector of the UW Medicine Sleep Center in Seattle. "On the other hand, that'd be a way to discriminate against people for whom this may be the only job they can get, or maybe the highest paying job they can get." The situation could get particularly dicey if, for example, genetic tests determined that the job would be bad for that worker's health and dissuaded an employer from hiring them because of associated costs of health care and absenteeism. It might even amplify age discrimination.

Younger workers tend to be more adaptable to shift work than older workers, which might explain why the younger Bruce Richardson could adjust to the twenty-eight-hour-day in Mammoth Cave and the older Nathaniel Kleitman could not. For one, circadian clocks in younger people already lean later and usually don't have as far to budge. We also become less phase tolerant with age. While both younger and older people can partially calibrate their circadian system to new sleeping and working hours, younger people cope more easily with the change. They tend to sleep better during the day and stay alert and perform better at night. Because we can expect the population of shift workers to parallel a progressively older society, this problem will become increasingly pressing in the years to come.

The next best option for a shift worker, at least one not assigned to a rotating schedule, might be to try to realign their rhythms permanently—essentially, to turn their biological night into day. That requires sticking to the same flipped schedule seven days a week, even during vacation time. This is rarely possible given the sun's insistence on shining during daylight hours and the average worker's assorted daytime obligations on days off. Experts told me that it can work for some people, like Brockman and other submariners underwater for months at a time. A nocturnal life may also be somewhat doable for certain destination gigs, such as in a mine or on an oil rig or at an Antarctic station. It may be achievable for a single, introverted person. And it could work for people like Finelli, whose partner is a night owl. Otherwise, a so-called compromise schedule is generally the best bet, said Helen Burgess, a sleep and circadian scientist at the University of Michigan.

The compromise entails nudging the circadian system a little bit nocturnal, to the point that the worker can shrink their swings in sleep and wake times across the week. In one study of shift workers, reported daytime drowsiness was less related to how long they slept and more contingent on when they slept. Shift workers who better synced their sleep and wake cycles with their circadian rhythm had lower daytime sleepiness, even if they slept fewer hours. One way to accomplish this, Burgess said, is for a shift worker to avoid exposure to light during their morning commute home and then promptly go to bed. If they can't time the drive before sunrise, she suggested, sunglasses can help so long as the glasses come off if the worker feels unsafely sleepy. Then, on off days, they could aim for a late bedtime and a late wake time to reduce the inconsistency. This is one situation when sleeping in until noon on the weekend might be encouraged. Research shows that these partially re-entrained workers sleep better and perform better than those not re-entrained at all. Still, even such a compromise may prove impossible for a lot of people.

I heard several other approaches in my conversations. Christopher Jung, the circadian scientist and avid whitewater rafter in Alaska, suggested that night shifters could work six days, take a day off, work another six days, and then take two weeks off. "It's essentially a full month of work within thirteen days," he told me. "That way, you can stay shifted to night and then shift back for the next two weeks." The result would be less flip-flopping.

Solutions become even harder to come by for workers on rotating shifts, including those Puget Sound Pilots. The typical strategy here, said Burgess, is to power through. To at least minimize harm, employers can reduce the duration of shifts and limit a worker's consecutive nights on the job. Because most human circadian clocks have a period slightly longer than twenty-four hours, an employer could take advantage of that natural delay and start a worker's shifts later with each rotation. For example, a worker might start one shift in the morning, the next in the early evening, and the next late at night. The slower the rotation, the better.

In almost every shift-work scenario, multiple tools can ease the burdens. Well-timed light can reset circadian clocks, and acutely boost alertness and performance—in compensation for working at the wrong circadian time. When necessary, caffeine, too, can enhance vigilance. To further mitigate the deleterious mental and physical health impacts, workers can restrict meals to the hours that their bodies think is daytime. And, at the end of the "day," a cool, quiet, and dark sleeping oasis is a must. Experts recommend eye masks and blackout blinds.

Finding the optimal approach is a constant and tricky balance between short-term safety and long-term health. And, as with most things circadian, it depends on the person and their circumstances. Olivia Walch, the mathematician who helped me with my bunker data, developed an app called Arcashift that aims to provide dynamic and tailored advice for shift workers. It collects personalized data predictive of internal time, and then advises shift workers when to eat, avoid light, or imbibe caffeine, to fight what she calls "massive circadian whiplash."

Walch envisions a future in which an advanced personalized system can read rhythms and light exposure and automatically adjust indoor lighting across day and night. "So anytime you have a disturbance to your schedule, your home just changes around you," she told me. Shift workers aren't the only people who could benefit. It could be used in workplaces and even travel environments where people are prone to circadian disruption. Such an integrated system could even be leveraged to manipulate or maintain rhythms for optimal athletic performance or medical treatment, such as ahead of a game, surgery, or an infusion.

Timeshifter, SleepSync, and other apps offer similar support for shift workers and advice for all of us in overcoming jet lag. The science continues to add to this toolbox. When switching time zones, we can prepare by steadily adjusting our sleep times in advance. Eating a big breakfast after reaching a destination might speed recovery. A strategically timed workout or supplemental melatonin could help, too.

Like light, melatonin has a phase response curve, meaning how intently your circadian system listens to a melatonin signal and how it responds to that signal will change across the day. It's all relative when your body naturally begins secreting melatonin. Melatonin supplementation early in the day, body clock time, can dial the clock back—helpful for most people traveling west. For eastward travel, taking melatonin in the evening will nudge the clock forward. And, of course, seeking and avoiding light at appropriate times will hasten acclimation, too. The rule of thumb for shift workers and globetrotters alike: exposure to light in the hours before the body's core temperature minimum, which falls around the midpoint of sleep, delays the clock; light in the hours after the minimum advances the clock. Light closest to that minimum, on either side, will produce the greatest pushes and pulls.

That is a fine line. Melatonin or light at the wrong time could reset the circadian clock in the opposite direction. I learned this lesson the hard way on my trip to Europe. I flew a red-eye from New York City to London and immediately upon arrival began walking the London streets with my luggage in tow as I searched for my Airbnb near Buckingham Palace. At that point, my internal time was probably around 3:00 a.m. eastern time, and my core body temperature was probably still on its way down to its minimum. Rather than advancing my clock, that light exposure probably delayed it—casting my internal time closer to central time than Greenwich mean time. I should have hidden from that morning light or put my blue-light blockers to use. At least I was right to seek out the sun on my return trip west.

My travels in both directions gave me a taste of what millions endure every day. I'm fortunate now to have the knowledge and tools to hopefully lessen the impacts next time, an easier feat for me than for shift workers burdened with chronic jet lag.

ERIKA GARCIA, THE NURSE IN MICHIGAN, TOLD ME THAT NO EMployer offered her help handling shift schedules. That glaring absence is typical for workplaces and a major point of frustration for experts like Philip Cheng, a clinical psychologist with Henry Ford Health in Michigan. "We're asking a large segment of our workforce to work nights, to provide essential services, but we're not providing any support," he said. Cheng wants to see more workplaces provide tips during orientation on things like communicating with family and friends about shift-work schedules, designing a bedroom for daytime sleep, and using light-blocking products and smartphone settings. After all, it's in the employer's best interest to maintain a healthy workforce.

Toward the same end, experts are encouraging investments in disease prevention and management: additional breast cancer and stroke screening, for example, and intensive behavioral counseling on nutrition, healthy eating behaviors, and increased physical activity. Federal policy could be another powerful way to improve working conditions and employee health. For truck drivers, stricter rules could level the playing field and reduce competitive pressures that keep trucks on the road day and night, said Ryan Olson, the occupational health scientist. There are caveats, of course. Fewer daily hours on the road, for example, would slow a long-haul trip and extend the days that a driver is away from home.

As society recognizes the true cost of our modern conveniences, including the sacrifices made by shift workers, what concessions might we make? Are we willing to give up late-night dining and gym time? How about waiting a little longer for a delivery? We can influence healthier work schedules for others with our own consumer behaviors. Still, even if we collectively decided to give up overnight deliveries and twenty-four-hour restaurants, gyms, grocery stores, and gas stations, shift work is not going away anytime soon. Certain round-the-clock professions will always be vital, such as firefighters, police officers, military personnel, ship operators, nuclear power plant workers, doctors, and nurses. In fact, we will need yet more health care workers on overnight shifts as the population ages. "It's a tough nut to crack," Burgess said of shift work. "At the end of the day, we shouldn't really be doing it."

OUR BACKWARD SCHOOL START TIMES MIGHT BE A SLIGHTLY EASIER nut to crack.

In the fall of 2016, morning bells at high schools in Seattle began ringing at 8:45 a.m.—about an hour later than they had the previous school year. Cindy Jatul, who then taught at Roosevelt High School, helped push for the change. It was a long fight. Eventually, with top scientists and doctors testifying before the Seattle Public Schools board, the revised schedule went through. Then, the benefits began to accrue. Before-and-after data collected from a sample of students wearing activity watches showed a gain of thirty-four minutes of sleep a night. Grades also improved and remained better in the years after the study, noted Jatul. Her daughter Ellen told me she felt better prepared for school with the later start time. She was in a happier mood, participated more in classes, and retained more information. Starting school later was a particular boon for students at lower-income Franklin High School, where the boost in sleep came alongside drops in absenteeism and tardiness. "That extra thirty-four minutes of sleep—that's huge," said A. J. Katzaroff, a Franklin High biology teacher. "I think they learn more and are more engaged in classes. It's so much more fun to teach kids that are awake." Many Seattle middle schools have now also delayed their start times.

Still, with stakes and tensions high, change can take time. In Seattle, it took nearly five years. The logistics of bus and sports schedules—complicated by a lack of athletic field lighting—were major sticking points and continued to be a point of contention after the change at Roosevelt High School. The concerns are legitimate. Darkness descends around 4:00 p.m. during the Seattle winter, limiting daylight hours for sports practices. The high school football coach reportedly rescheduled his team's practices to before school. Other coaches did the same. Horacio de la Iglesia, the University of Washington circadian biologist, led the before-and-after study, titled "Sleepmore in Seattle." He argued that this defeated the purpose of the change: "We know that sleep-deprived kids don't perform well in sports. They also get injured more easily." And they can be extra prone to injury in the early morning, when their muscles have yet to warm up.

While the older kids got a later start, the morning bell began ringing earlier at most Seattle elementary schools. Jatul was in favor of that move, too. "Elementary-aged kids are bouncing off the wall at 7:00 a.m.," she said. The revised bell time also balanced out the bus schedules.

Seattle was not the first place to change start times and see benefits. School districts in Minnesota delayed high school bells in the 1990s and found the average student began scoring higher on standardized tests and sleeping about an hour longer on school nights. A study conducted after schools in Wake County, North Carolina, delayed start times in the early 2000s concluded that a morning bell one hour later led to a gain of at least one percentage point in reading test scores and two percentage points in math test scores for the average student, with stronger benefits for students at the lower end of the distribution. And shifting the start time in 2017 from 7:10 a.m. to 8:20 a.m. for high school students in Cherry Creek, a neighborhood of Denver, netted them an additional forty-five minutes of sleep.

In 2014, the American Academy of Pediatrics and the American

Academy of Sleep Medicine issued policy statements against middle and high schools starting before 8:30 a.m. The American Medical Association soon recommended the same. The state of California took the advice and, in 2022, became the first state in the nation to require public high schools to begin no earlier than 8:30 a.m. and public middle schools to start no earlier than 8:00 a.m. Other states, including Florida, have begun following suit.

Based on estimates of improved academic performance and reduced car crashes, a RAND analysis concluded that later school start times could prove an economic and public health boon as well. "A small change [in school start time] could result in big economic benefits over a short period of time for the U.S.," stated one researcher. "In fact, the level of benefit and period of time it would take to recoup the costs from the policy change is unprecedented in economic terms." Organizations such as the nonprofit Start School Later continue to advocate for "healthy, safe, equitable school hours" and offer templates for other schools to follow.

The movement is global. Some high schools in the Netherlands and Germany now give students options for when they come to school. Core subjects fill the middle of the day, between around 10:00 a.m. and 2:00 p.m. Students can then choose to take their electives in the morning or afternoon.

Even if a middle or high school cannot delay or flex its starting bell, it could still help students by not giving early morning exams, just as workplaces might delay morning meetings. Early afternoon is ideal for administering tests, Kring said, because that is "when we find the most equality between the chronotypes." Denni Tommasi, the economist in Italy, came to that same conclusion after seeing those unexpected patterns in test scores. Based on his data, he told me, College Board, the administrator of the SAT, should move its traditional 8:30 a.m. to 9:00 a.m. start times to later in the day. After the test went digital in fall 2023, however, start times moved even earlier—to between 8:15 a.m. and 8:45 a.m.

In his study of college student test scores, Tommasi also found a far greater morning disadvantage in January compared with May and June, presumably due to a later sunrise and even less daylight reaching students' eyes before morning exams. Another study hinted at detrimental effects of daylight saving time on SAT scores. The state of Indiana, which has historically observed a hodgepodge of social clock times, served as a natural experiment. Researchers found a sixteen-point lower average for students who lived in parts of Indiana observing seasonal daylight saving time versus permanent standard time. When they broke scores down by socioeconomic status, the average daylight saving time detriment for students from the poorest households was forty-nine points.

CONCURRENT DELIBERATIONS CAN BE HEARD ACROSS THE US AND Europe: Should we delay school start times? Should we advance clocks to stay year round on daylight saving time? The two ideas are contradictory in many ways. A move to permanent daylight saving time would undo gains made by delaying school start times an hour: "Then they would need to start two hours later," Jay Pea told the Kansas House Committee on Federal and State Affairs in February 2021. Pea, a former software engineer and the founder of the nonprofit campaign Save Standard Time, was giving virtual testimony against a bill that urged the US Congress to extend daylight saving time throughout the year for the nation.

Twice a year, much of the Western world still changes their clocks. And, about twice a year, debate heats back up over whether to ditch the antiquated practice. Public polls suggest that most people would be more than happy to stop doing it. But there is no consensus on which way to go: Should we lock our clocks forward or back?

I met Pea for lunch at the corner of Division and Henry Adams Streets in San Francisco to discuss his campaign. As we approached each other, he lifted his gaze over my head. I watched him look down at his left wrist, then up over my head again. An odd way to greet someone for the first time. I then turned and followed his line of sight to a clock tower on the Sobel Building across the street. "What time is that clock on?" he asked aloud. As I squinted up at the clock, I saw it read just after 1:00 p.m. He showed me his watch, which read 11:02 a.m. My Fitbit split the difference: 12:02 p.m.

Pea, I learned, hasn't swayed from a decree he made in 1988, when he was ten years old, to keep his watch permanently set to standard time. Today, a sun-powered analog watch on Pea's left wrist shows standard time, all the time. Every day, he goes to bed when his watch reads around 9:30 p.m. and he gets up eight or nine hours later—almost always without an alarm. According to Pea, the discrepancy between the time on his watch and the social time has only occasionally gotten him into trouble. There was that one time during college, just after the switch to daylight saving time, when he arrived late for a haircut. If that's the worst conflict he's faced, I thought, he's doing just fine. When we met, the local time displayed on his iPhone added social insurance. He has since relocated to Arizona. The state observes standard time year round, which he told me was a leading factor in his decision to move there from San Francisco. Now, his wristwatch never deviates from the local social clock.

Over the last few years, Pea has been on a quest to get everyone, including other states and countries, to join him in ditching daylight saving time and adopting permanent standard time. He faces an uphill climb to convince the masses and the policymakers, many of whom are pushing to make daylight saving time year round. His side, however, has the backing of science and history. Pea has testified in several states, from Oregon to New Hampshire, and has worked with circadian scientists to clarify common misconceptions. If you were to ask most Americans whether they would willingly set their alarm clocks an hour earlier in the winter to get up and go to school or work in the dark, on icy roads, sleepy all the way, "they would probably say no," Pea states in a Save Standard Time campaign video.

Pea highlights the reasons to abolish daylight saving time—from the accidents and other acute effects of the lost hour of sleep to issues that linger long after the change. He reminds lawmakers that the US already experimented with permanent daylight saving time—multiple times. It repeatedly failed. The United Kingdom tried it beginning in 1968, only to eliminate it after three years. And Russia adopted it in 2011 but also changed course after three years in favor of permanent standard time or "winter time," as it's called there. "Do not repeat bad history," Pea said during his virtual testimony to the Kansas committee.

Pea pointed out, and I would agree based on the conversations I've incited with friends on the topic, that many people who say they prefer daylight saving time really just prefer having more daylight after school or work. I had always been with that majority, focused on that bonus late-in-the-day light. I am probably not the only one who has associated daylight saving time with the happy days of summer. But I had also been unaware of the relevant facts and figures. For instance, I didn't know that Russia saw a rise in the rates of adolescent winter depression when it switched from seasonal to permanent daylight saving time in 2011. Or that rates dropped again after the country changed in 2014 to permanent standard time. Morning light, as we know, is a choice treatment for seasonal affective disorder.

Ultimately, no manipulation of the clock can lengthen the short days of winter to match those blissfully long summer days. "What we need is more flexible schedules," Pea said. As we left the restaurant, he and I both gazed across the street, up at the Sobel Building clock tower—it still read just after 1:00 p.m. The clock was broken.

DOZENS OF US STATES ARE ACTIVELY DISCUSSING LEGISLATION TO stop the biannual tug-of-war between our biological and alarm clocks. Federal law already permits states to adopt permanent standard time. For a state to switch to permanent daylight saving time, however, Congress would have to change the law. As I write this, Washington is one of nineteen states that have enacted legislation to allow them to adopt year-round daylight saving time should Congress make the move.

Politicians from Florida, widely considered the nation's top state for golf, are leading the charge. In 2023, Senator Marco Rubio reintroduced a bipartisan bill, the Sunshine Protection Act, which would make daylight saving time year round for the whole nation. Representative Vern Buchanan introduced a companion bill in the House. Support for both bills grew after their first iterations in 2018. By 2022, the House bill had forty-eight cosponsors. Permanent daylight saving time threatened to become law that year, with a unanimous vote by the Senate. The bill died in the House.

The pro–daylight saving time side of the debate also had its citizen face during that time: Scott Yates, a fiftysomething Colorado-based writer, entrepreneur, and creator of the #LocktheClock movement. He testified during the Kansas hearing as well, telling the committee that their daylight saving time resolution was "perfect" and urged its passing, as he'd done for several other states. He highlighted benefits of more day-end light—greater opportunities for kids to exercise, lower rates of crime. There's no doubt that schools lacking athletic field lighting might have an easier time with after school sports. However, the data are arguably mixed. Research that appears to support permanent daylight saving time is often based on effects detected during clock change transitions and may not reflect year-round, long-term impacts. For example, short-term evidence shows a drop in certain crimes and nighttime car crashes with the change to daylight saving time. But standard time supporters point to evidence that these are offset by more morning collisions and other factors that affect longer-term rates of accidents and crime, such as attention lapses and mood changes due to sleep deprivation and circadian disruption.

When we spoke in 2021, Yates conveyed his conviction that "TV people" want it to be "dark earlier so that everybody is inside watching sports." He added, "I'm not a scientist. I'm not a researcher. I just try to help whoever is trying to do whatever they can do so that we stop changing the clock twice a year." You'd think he and Pea could be on the same team. Rather, they couldn't have been at more bitter odds. In the months before we spoke, the two feuded on Twitter and via blog posts. Yates reportedly leveled multiple accusations at Pea, suggesting that he used a fake name and hinting that the TV industry funded his campaign. Yates has since stepped "behind the scenes" on the daylight saving time advocacy, according to his website. He unsuccessfully ran for election to the US House of Representatives in 2022.

The once mundane question, "What time is it?" has become loaded. Emotions are high. Yet only one side of the debate has the broad support of health scientists and medical doctors. In 2020, the American Academy of Sleep Medicine—among the organizations also advocating for later school start times—issued a position statement in favor of permanent year-round standard time, suggesting it "aligns best with human circadian biology and provides distinct benefits for public health and safety." Their position, which has been endorsed by more than twenty health, safety, science, and education organizations, gained traction in early 2024. Several states impatient with the lack of action from Congress on permanent daylight saving time, and desperate to stop clock changes, resorted to consideration of permanent standard time.

There might be environmental perks with that option, too. While saving energy was an early argument for daylight saving time, any small drops in electricity use for lighting at night appears to be offset by increases in gasoline use, as well as evening air-conditioning and morning heating that come with a later sunrise and sunset. Arizona found energy consumption soared when the state initially followed daylight saving time in 1967, before opting for standard time. More recent natural experiments in Australia and in the state of Indiana also failed to find an overall energy savings with daylight saving time. One study actually saw a slight rise in electricity use. The balance might shift further as the energy efficiency of lighting continues to improve, and as people learn they don't need to use so much of it after the sun goes down.

It seems the whole world is taking hard looks at their social clocks—

at least nations at higher latitudes. Alberta, Canada, narrowly voted down a referendum in October 2021 for full-time daylight saving time. Voters were asked if they wanted to adopt "summer hours." The referendum's wording did not point out that making the permanent switch would bring weeks in the winter when the sun wouldn't rise until well after 10:00 a.m. across much of the province. The European Union, too, launched an effort in 2018 to stop fiddling with clocks in the spring and fall. Progress stalled, likely due to pressing matters, such as Brexit and the war in Ukraine. There was also a lack of consensus among nations on the choice between permanent summertime and wintertime, their terms for daylight saving and standard times. Again, scientists fervently warn against that move, just as they applauded Mexico's 2022 decision to end its seasonal observance of daylight saving time and restore standard time year round, or what the country's health secretary called "God's clock." Iran, too, went to permanent standard time in 2022.

At least one expert suggests that springing forward and falling back, despite its drawbacks, may remain the lesser evil. David Prerau, author of the book *Seize the Daylight*, has called the current system a "sensible compromise." He advocates simply softening the impacts of the switches. For example, he told me, a public service campaign could remind people several days in advance that a spring forward or fall back is coming. And it could recommend strategies akin to preparing for jet lag, like getting more sleep and slowly adjusting bedtimes ahead of each clock change.

I can see his point. I personally don't need the sun rising at 4:11 a.m. in mid-June to get decent morning daylight. Yet that's how the social and sun clocks would line up in high-latitude Seattle under permanent standard time. Too often summer morning sunlight already sneaks around my blinds and wakes me up. The issue is those late winter sunrises that would come with permanent daylight saving time. Yet I'm also convinced that repeatedly borrowing and returning that morning hour of daylight is legitimately bad for our bodies, too.

Optimally setting the social clock is a formidable challenge. There's no perfect solution. Thankfully, society has other tools at hand. Why not keep clocks on standard time and add slack to our schedules? Again, there is precedent. More than two hundred years ago, the Spanish parliament regulated that the president would "open sessions at ten from October 1 to April 30 and nine from May 1 to September 30."

A FEW SCHOLARS AND TECHIES ARE EVEN MORE AMBITIOUS IN THEIR vision for the future of social clocks. Today's digital world has become increasingly interconnected, significantly more so than in the 1880s when the railroads insisted on simplifying timetables with standard time zones. Proponents argue that running the planet on one time would create far fewer timekeeping problems across the globe. They recommend using Greenwich mean time or coordinated universal time (UTC), which is based on the cycles of atomic clocks rather than the rotation of the Earth. People could still live their days by the sun, their clock just might read noon in the middle of the night. The International Space Station, financial institutes, and the US military already use UTC. Whenever Kevin Brockman's submarine traveled a decent distance, he told me, his crew would swap local time for UTC, or what the military calls "Zulu time." At that point, he said he lost all sense of time: "It could be midnight based on your clocks, but it's, you know, 6:00 a.m. wherever you actually are."

In 1976, acclaimed science fiction writer Arthur C. Clarke hinted at the idea in the same interview in which he predicted the smartwatch: "In the global village of the future, it will be like living in one small town where at any time about a third of your friends are asleep but you won't even know which third. So, we may have to abolish time zones completely and all go on a common time, the same time for everybody, which will cause all sorts of problems." The idea of whittling down time zones to one came up again in 1998: "Cyberspace has no seasons and no night and day," said Nicholas Negroponte, director of MIT's Media Lab. Back then, he was introducing preproduction models of a Swatch watch tuned to "internet time," based on a new global meridian centered at Swatch's factory in Biel, Switzerland.

Without imposed time zones, the old "nine-to-five" would lose meaning. Perhaps that would ease setting up a new time architecture for work, school, and play. A Norwegian island appeared to take this concept to the extreme in 2019, when residents declared Sommarøy the world's first time-free zone. It turned out to be a publicity stunt to promote tourism.

What if we instead allowed every town to once again set their own clocks to the sun? Martha Merrow of LMU Munich would prefer to see us reestablish these hyperlocal time zones, with our social clocks following the sun's trek across the sky. The day would again be spread out symmetrically from high noon. "We don't have to have time zones anymore," said Merrow. "Sun time is much more intuitive." A watch app called Circa Solar allies with that philosophy. The simple pie-like display shows only white between sunrise and sunset, black for nighttime, and shades of gray in between for dawn and dusk. A line indicates the current location of the sun in that twenty-four-hour cycle—with the relative black-and-white portions of the daily pie shrinking or enlarging with the seasons.

More practical, or at least more immediately doable, would be to redraw or add time zones in parts of the world. China could certainly use a few. As could India. Spain would benefit from switching from central European time to Greenwich mean time, a seemingly obvious move given it sits longitudinally in the middle of the latter zone. Till Roenneberg and colleagues have proposed one comprehensive solution that involves a combination of "obliterating" daylight saving time in favor of permanent standard time and reassigning countries and regions to their sun-clock-based time zones.

Whether or not we do away with, readjust, or radically raise the number of time zones, and even if we fail to unite on permanent standard or daylight saving time for the clocks on our walls, we still have the power to live and work better by the clocks within us. We can revise and relax when we require employees and students to arrive in the mornings. We can each do our best to regularly rise and rest relative to the sunrise and the sunset. Overall, we can reorient schedules around natural cycles rather than an outdated social time architecture. At least one circadian group is developing a curriculum to instill this message to the next generation. "All children learn that it is absolutely normal to use an alarm clock, that it is absolutely normal to feel tired and groggy in the morning," said Thomas Kantermann, a chronobiologist at the FOM University in Essen, Germany. "This is what we teach them. Why?" The materials may be just as useful for adults, whose unnatural daily patterns have been reinforced over decades. "It's very difficult to change habits," Kantermann added. "This is why it's easy for pharmaceutical companies: it's easier for us to take a pill once a day and not think about anything." Or it's easier to stop by Starbucks on the way to work for that venti latte.

Small changes, such as delaying clock-in times by an hour or two, could pay valuable dividends for health and productivity. Again, company leaders yield a lot of power to help their employees. AbbVie in Norway and Kring's other clients were ahead of the curve in recognizing that the schedules required for a farm, or an assembly line, rarely apply to today's workforce. Few good things came out of the COVID-19 pandemic. But one of them was an eye-opening lesson for many more companies and their employees: the flexed work hours during the pandemic offered a glimpse of what a more sun-synced life could look and feel like. Studies found that people, especially night owls, tended to get more sleep and endured less social jet lag as school and work schedules were relaxed, though some data suggested a slight drop in sleep quality—likely attributable to the stresses of the unprecedented time. The pandemic "has taken a huge magnifying glass to the things that are still very wrong in our society," Roenneberg told me. And it has opened up the possibility of a radical new workday for traditional nine-to-fivers.

The schedules of our daily lives can be complex. It may be difficult to change one component in isolation. Better might be to pull all the pieces together at the same time, in the same place. That was the idea behind the world's first ChronoCity.

MICHAEL WIEDEN'S FIRST ATTEMPT TO ROLL OUT THE SCHEME—A plan that included tinkering with work and school schedules, artificial lights, mealtimes, and even the check-in times of hotels—was in the Bavarian spa town of Bad Kissingen. It was 2012. Unfortunately, it was bad timing. While Wieden, a business developer, and his scientific team initially had widespread support, practical challenges and partisan politics quickly quelled the novel effort. Circadian science hadn't yet risen into the public consciousness. What's more, much of the town's workforce and many of its students commute in from outlying rural villages, often by bus. Revising bus schedules to fit new work and school times posed a problem.

A few of Wieden's initial makeovers did stick, such as later test times and circadian-stimulating lighting at Jack-Steinberger-Gymnasium, the local high school. But the greatest gain from the effort, he told me, was progress in the public's awareness of biological rhythms. The unconventional experiment caught the attention of people around the world. The ChronoCity project in Bad Kissingen may have failed. "But it's not the idea that has failed," said Kantermann, who worked with Wieden on the project. "Literally, it takes time."

Thanks to the explosion in circadian science over the last several years, and growing public interest, the ambitious pursuit of weaving chronobiology into the fabric of society now has a legitimate chance. The data point to an increasingly clear message: our clocks are off, and it's time we set them right. "This is what the science tells us. So, what do we do about it?" said Kantermann. "The bottom line is, we need a little revolution."

Roenneberg agreed: "All you have to do is get enough people together who have the guts to not be bourgeois." He had advised Bad Kissingen to declare it was sticking to standard time. He told them that doing so would attract global attention and get the ball rolling. But leaders were afraid of lawsuits from people who might end up filing something at the wrong time, missing a grant or other deadline, he told me. "They pussy-footed out," he said. "They missed their chance."

The ultimate goal for Wieden, Roenneberg, Merrow, and many experts I spoke with is to eliminate the need for man-made alarm clocks. For that to have a chance, our internal alarm clocks need a lot more of a rare commodity: the right light at the right time. Thankfully, as we'll soon see, recent scientific and technological breakthroughs could usher us into a new age of enlightenment—significantly brightening our days and darkening our nights.

10

Let There Be Light, and Dark

The glowing white letters of the Malmö Arena are hard to miss when you emerge from the train station of this coastal city in southern Sweden. I'd arrived from Copenhagen on a drizzly October day. The lack of sunshine bothered me less than usual, though, as I had been assured that a convincing replica awaited me straight ahead—across the rain-soaked courtyard and inside the arena doors.

This is where the Malmö Redhawks professional hockey team plays its games. As it logically follows but hadn't yet hit me, this is also where the players take off their sweaty equipment—skates, helmets, pads, gloves. I may have come to view some fancy lighting, but the first thing to impress me was the fantastic stench. And then I saw the light. An array of more than a dozen rectangular LED fixtures spotted the ceiling of the Redhawks' locker room, the focal point of the tour led by Freddie Sjögren, head of performance for the team. Sjögren wore a black T-shirt printed with the Redhawks' fierce bird logo. Open lockers flanked three sides of the room—team bags, skates, and hockey sticks spilled onto red carpet emblazoned with the same logo. On the fourth wall hung a pair of whiteboards shaped like hockey rinks and a large digital clock.

Easy to miss in this busy, colorful scene was a small square control panel mounted between the door and a line of lockers. As Sjögren showed me, a press of one of the inconspicuous white buttons on the panel could dim the lights, turn on a preset scene, or trigger an automated setting that adjusted the intensity and spectrum of light fourteen thousand times throughout the day—an attempt to closely emulate the subtle changes in natural light as the sun arcs across the sky. Rather than a typical gray October day in Sweden, BrainLit, the company that designed and installed these tunable LEDs, chose to approximate an April day. The idea was for bright bluish light to energize players before games and training, softer orangish light to wind them down after, and the whole sequence to keep their circadian clocks consistently ticking on time.

As we walked on through the team's training room and weight room, where the novel lights also beamed and bounced, Sjögren pointed out the obvious: "We don't have windows." Players and staff spend much of their days in the dark depths of the stadium. And at such a high northern European latitude, days during the hockey season can be short. The team often arrives in the morning before the sun rises; they often leave well after sunset.

We continued the full-sensory tour down a tunnel into the chilly ice arena. Although the massive space was light enough that I could see my own breath, it felt extremely dim compared with the rest of the team's facilities. There were no circadian-stimulating LEDs here, at least not yet. "That would be awesome," said Sjögren. After a few minutes gazing at the rows of red seats leading up from the advertisement-filled boards that formed the boundary of the rink, the chill started setting in. Thankfully, Sjögren soon ushered us back down through the tunnel for a closer look at the locker room lights. This time, with his help, I noticed the old fluorescent fixtures remained in the ceiling, interspaced with the new LEDs. Sjögren switched on the old lights and then the new lights, back and forth, several times. The difference was striking. The latter lights made the space and its colors pop. "It's more like the sun," said Emil Sylvegård, who played left wing for the team.

Sylvegård and I stood in the locker room for a few minutes, soaking up that artificial sun. He told me he's felt more energized since BrainLit began installing the lights about three years ago. It's now easier for him to focus during the day and to fall asleep at night. "I get more deep sleep," he said. "Normally, after a game, I'm all pumped up." Sylvegård added that he thinks he's suffering fewer injuries, too. I heard similar feedback from Anna Milstam at the Lindeborg school, less than two miles away. Milstam's classroom was among BrainLit's first circadian lighting installations. She and her students shared several stories about perceived improvements in their sleep, alertness, and performance at school. Of course, there's no way to prove the lighting deserves credit for any of the athletic or academic gains. And the placebo effect is real. Still, it seemed worthy of further exploration.

The Redhawks, who play in the professional Swedish Hockey League, had lost in the arena two days before my visit, 5–2. I had been warned not to mention the game to any coaches or players. The day after my visit, they won 6–1. That was a road game. When I later saw that score from my Copenhagen Airbnb, I remembered something else Sylvegård had told me: "We're outside more when we travel." While going between cities, players get more natural daylight. The real deal. And given that they play all of their games within a relatively small country, they didn't need to worry about jet lag.

Months later, after the hockey season ended, I looked up the final league standings. The Redhawks had placed last—although they did at least win a playoff that kept them from getting relegated to a lower league. Most of their victories were away games. Even if the new lighting gave the team a slight edge, it may easily have been overshadowed by factors like the team's talent and the unsurpassed power of the sun.

I'D COME FULL CIRCLE. MY JOURNEY INTO CIRCADIAN RHYTHMS began unexpectedly back in 2014 with that visit to the locker room of my hometown Major League Baseball team. There, too, preset LEDs provided stimulating pregame light and calming postgame light. "When you're in sports, you're looking for every advantage you can get to improve performance," the Seattle Mariners' former facilities manager, Scott Jenkins, told me in 2018. Several football, baseball, basketball, hockey, and other sports teams have followed the lead of the Mariners and additional early adopters. When we spoke, Jenkins had just installed tunable LED lighting for the Atlanta Falcons, a National Football League team. The Tottenham Hotspur Football Club in London joined the bandwagon more recently. (That's European football.) Meanwhile, the Mariners have remained mediocre. I'll still blame the sleep and circadian disruption from their unfair travel schedule.

Sports represent a small slice of the circadian lighting market. And BrainLit is just one of a growing number of businesses around the world promoting the products to schools, offices, hospitals, nursing homes, and more. "There are only thirty-two NFL teams. There are only thirty MLB teams," Jenkins had told me. "But think about how many schools there are, and how many health care facilities there are. To be able to use light to benefit people's learning and health—there's a huge opportunity there." According to proponents, many typically windowless workplaces, such as call centers, warehouses, and pharmacies, are also prime candidates. Take, for example, the Malmö Arena kitchen. BrainLit had installed tunable LEDs there, too, a couple of months before my visit. I took a peek before leaving the arena. The kitchen staff told me that, before the lighting arrived, they would flee the darkness and hover around windows along the arena's perimeter during lunch. It took some adjustment to handle the brightness, but they seemed happy with their new lights. "It's much better now," one woman told me. "We're otherwise in a cave."

Doctors have long prescribed light boxes and related paraphernalia for SAD and other forms of depression. But it's only now, company leaders and other proponents say—amid innovations in LED technology, amid calls for more energy-efficient lighting, and amid a renaissance in scientific understanding of the biological impacts of light—that a revolution driven by more thoughtful lighting infrastructure has begun to unfold.

As we've learned, exposure to bright, blue-rich white light during the day and to softer, amber hues at night helps restore the human body's deeply ingrained, physiological drumbeat. By tweaking the color, intensity, and timing of artificial lighting, tunable LED systems could, at least in theory, reinfuse our predominantly indoor lives with some of that missing light-dark contrast—a distinction that standard, non-tunable LEDs have further depleted for many of us. That reinfusion, in turn, could help keep our clocks in sync, track seasonal changes, improve our alertness, boost our memory and mood, and fend off at least some of those troubling circadian-related health problems. Still, many experts temper their enthusiasm. The Redhawks' losing record despite their fancy lights justifies some hesitancy. There may be an opportunity for widespread benefits, but without a stronger scientific foundation, as well as supporting standards and regulation, there's also opportunity for not-so-great and potentially harmful products to enter the market. Can the science support all the claims? And how did this wild west of swanky lights arise in the first place?

LEDs have been with us for a while, ever since those Nobel Prize–winning scientists came up with that crucial missing blue diode to make white light. But only with newer tunable technologies can LEDs be programmed to shine the right light at the right time. Products now allow users to dial up the intensity and concentration of wavelengths, say to around 480 nanometers early in the day, and then dial both down at the end of the day. No longer are we beholden to the on-off switch. Lights can also be programmed to ramp up brightness slowly to counter any initial aversion. Indoors, intense light can at first appear harsh even when weaker than the light of a cloudy day outside. It's about that contrast.

Circadian lighting systems debuted around 2013. One of the first high-profile installations came in 2016, aboard the International Space Station. The system has since been upgraded to even better replicate the variable natural lighting on Earth. For the rest of us on the ground, the higher price tags for tunable systems compared with standard LED fixtures has impeded growth from the start. Energy-efficiency concerns further slow the pace.

As you may remember, traditional energy-efficiency standards lean on metrics like lux and lumens that are based solely on the responses of rods and cones and aim to illuminate spaces solely for functionality and safety. Building designers have had little incentive to consider the nonvisual effects of light. For a long time, no one could even agree on *how* to measure those other impacts. Finally, in the late 2010s, the Austria-based International Commission on Illumination recommended an international standard metric for measuring circadian-stimulating light: melanopic EDI. That became the metric used in the 2022 lighting design recommendations. Industry has now begun to adopt the concept. UL Solutions, which provides safety and performance testing and certification, and the International WELL Building Institute, which administers healthy building ratings, include both circadian lighting and daylight exposure in their latest guidelines for offices. Experts told me that they anticipate the U.S. Green Building Council's LEED certification program to follow suit.

Momentum is building. Circadian lighting products continue to grow in sophistication and drop in cost, sparking greater demand. The market for circadian lighting surpassed $1 billion in 2020. Analysts expect it to reach $5.5 billion by 2027. Installations can already be found in hotel rooms, aboard high-end commercial airplanes, and on commuter trains. I've now seen the technology in multiple schools, offices, a nursing home, a hospital, and two locker rooms. I watched a kindergarten teacher adjust her tunable classroom lights to energize and focus her students, and then later to calm them down. I even visited a café in Manhattan that serves blue blasts of light, in addition to Green Boost juice—Honeybrains advertises itself as "a happy place to nourish your brain."

There is little question that the study of human interaction with light is now in its heyday, and that the implications for our hopelessly indoor lives could be significant. We may not know everything yet, experts told me, but we know enough about the benefits of circadian lighting—and the downsides of current lighting—to put the ideas into practice sooner rather than later.

ONE OF MY FIRST ENCOUNTERS WITH CIRCADIAN LIGHTING, AFTER that initial introduction in the Mariners' locker room, was in 2017, when I visited the headquarters of Vulcan Inc., an investment and philanthropic firm. The offices spanned a few floors of a high-rise building on the south end of downtown Seattle, within eyeshot of the Mariners' stadium. I met Mountain Love in a tenth-floor common room. Straight out the north-facing window, he and I could see the Space Needle standing tall yet dwarfed between buildings and cranes of the fast-growing city. To the left, the waters of Puget Sound glistened under partly cloudy skies. To the right, about a block away, we could see a century-old local business that's long helped illuminate this skyline: Seattle Lighting.

Love, then a business process architect for Vulcan, had landed in the Emerald City eighteen years prior from Colorado, a far sunnier locale. "When I first moved here, I was really unhappy," he told me. Two years into his job at Vulcan, Love still struggled with sleep and focus. He needed frequent doses of caffeine to get through the workday. "Thank God there was a Starbucks downstairs," he told me. But Love's outlook brightened in 2014 when company executives asked for assistance in investigating the benefits of circadian lighting. A tunable LED soon appeared above his desk. In words that would become familiar to me, he stated the circadian lighting refrain: "I immediately started to feel better."

Vulcan's lighting retrofit was reportedly the first large-scale installation of tunable LEDs in a US office building. The motivations sounded much the same as those for flexing work schedules: "Your people are where most of the cost is," Cody Crawford, the head of facilities and operations at Vulcan, told me. His office was down the hall and past a photo of the burning sun from Love's desk. Crawford emphasized the importance of educating Vulcan employees about the lights. "People in general don't like change. They may not realize that the lack of light is making them tired and groggy," he said. "You almost have to force them to try it for more than an hour." Aside from an initial burst of alertness and energy, blue light's additional short- and long-term impacts can go consciously unnoticed. Crawford also couldn't be sure how the new lighting affected Vulcan's employees, or its bottom line. It's hard to measure productivity "unless you're a manufacturing facility," he said.

Studies elsewhere have taken on the challenging task of deciphering the effects. Many have reported improved alertness, performance, mood, and sleep quality with higher melanopic EDI during the daytime. Brigham and Women's Steve Lockley and colleagues found that people exposed to blue-rich light reacted faster and had fewer lapses in attention than people exposed to longer-wavelength light. Another small study found an hour dose of 40-lux blue light superior to the caffeine common in a sixteen-ounce cup of coffee in raising certain measures of cognitive function and alertness.

More research is needed to objectively measure things like productivity, accidents, sick days, and health outcomes. But that data, too, is beginning to come in. A report commissioned by LightingEurope, an industry trade group, calculated office productivity gains equivalent to about two more hours per month, a decrease in sick days of 1 percent, and an increase in employment duration of one year with tunable LED lighting. And in studies led by Mariana Figueiro, then at Rensselaer Polytechnic Institute and now director of the Light and Health Research Center at the Icahn School of Medicine at Mount Sinai in New York, employees who received high levels of circadian-stimulating light during the morning had more stable circadian rhythms, less depression, and greater sleep quality compared with employees receiving lower levels. The benefits of improved sleep alone could be substantial. Insufficient sleep is estimated to cost the US economy more than $400 billion a year due to absenteeism, accidents, and reduced productivity. Studies of students from preschool through college have reached similar conclusions.

STILL, AS COMPANIES CONTINUE TO PROMOTE PRODUCTS THAT PROMise to improve our sleep, smarts, and speed, and to help us to smile more, some scientists worry about the potential for harm. More question whether circadian lighting products can live up to the hype.

Anna Wirz-Justice, of the University of Basel, joins other experts wary that enthusiasm has outpaced the science and the regulation. "The lighting industry has jumped on circadian lighting," she told me. "Yet most of these devices have never been compared in any kind of clinical trial." Regulation by the US Food and Drug Administration kicks in when a product is labeled a "medical device." Lights need not be. Companies can add blue wavelengths and call their lights circadian as long as they still meet the safety regulations set for traditional lamps and don't make medical claims they can't back up.

If products marketed as circadian lighting aren't "under the scrutiny of a medical device," then they should carry a warning label, said Deborah Burnett, a lighting consultant based in California. A consensus survey of 248 leading scientists agreed. In a 2023 paper, they recommended warning labels on blue-enriched LED lights that state the products may be "harmful if used at night." However, the scientists also agreed that the introduction of tunable LED lights was a good thing; the light's blue content just needs to be altered across day and night. As Burnett put it: Lighting manufacturers need to consider the "complete balance of light" across the twenty-four-hour day.

The risks to the brain and body from the wrong light at the wrong time are not always obvious, said Burnett, echoing what other scientists—and Vulcan's Crawford—had noted about the subtleties inherent with both lighting and circadian health. If you mess up your circadian clocks, it's not the same as putting your hand on a hot stove and getting a blister.

Rather, it is a long, slow burn. "It's like a frog in boiling water," Burnett told me. "By the time the frog goes, 'I better jump out,' it's already too late." Beyond the health hazards, Burnett warned that companies can easily creep into the promotion of "gimmicks" and "snake oil."

She described a cautionary case from the 1980s. The manufacturer of Vita-Lite, a full-spectrum fluorescent lamp designed to simulate natural sunlight, claimed in its promotional literature that it "helps fight winter blues," along with other health benefits. The FDA countered, noting that studies published at the time suggested "no advantage to full spectrum lights like Vita-Lite compared to ordinary cool-white fluorescent bulbs or standing in the sun." The lamp aimed to mimic the sun, not outperform it. In any case, Vita-Lite's claims also included fewer dental cavities in hamsters and increased sexual potency in male turkeys. The FDA cautioned that this "might make unwary consumers think they could gain similar dental or sexual dividends," and ultimately charged the manufacturer with health fraud. "Much more research is required before firms could justify medical claims for non-prescription use of full-spectrum lamps," the FDA said.

Was Vita-Lite a fraud or just ahead of its time? Science has progressed substantially over the last four decades. It's even clearer now that we need to brighten our days and darken our nights. But open questions about optimal wavelengths, intensities, timing, and exposure durations remain—especially for humans outside of lab settings. In search of answers, one team at Oregon State University is constructing a futuristic trailer that will house volunteers going about their near-normal daily lives. The researchers will be able to adjust how much artificial light is emitted by LEDs and how much natural light penetrates the trailer through high-tech windows that automatically tint based on cloud cover and the angle of the sun. They plan to measure various effects on volunteers, such as sleep and alertness. Other scientists are leveraging functional MRI technology to understand the brain's response to light. Still others are continuing to discern more details of our complex nonvisual photoreception circuitry via animal models.

As neuroscientists, biologists, sleep doctors, and architects worldwide investigate these questions, the answers tend to grow more elusive and multiply. Whether exposure to light will help or hinder someone's rhythms comes down, primarily, to getting the right color at the right intensity at the right time and for the right length of time. It might even depend on the angle of the light. Figueiro is finding that light aimed around the nose, where it can directly strike both eyes, may most effectively stimulate the circadian system. A ceiling light right above the head or a lamp set peripherally to a person's line of sight may miss the target. This is already a lot of dials to turn. Plus, we know the optimal recipe can vary by age and other differences between us.

Further complicating this research is the difficulty of measuring our true hour-to-hour light exposure. The direction a person is looking, and even how they tilt their head, can also drastically change how much light reaches their eyes. What's more, standard recommendations for lighting in buildings are based on how light illuminates a floor, desk, or other flat surface, rather than the vertical plane where—as Figueiro explained—the light actually hits the eye. Effective circadian light exposure depends on the placement of a fixture and how photons emerge from that fixture and bounce around a room. The Department of Energy estimated about a fivefold variation in circadian-stimulating light between one hundred forty-two workstations under the same electric lighting. These factors help explain why Shadab Rahman, the Harvard sleep scientist, calculated that paltry 50 to 100 lux in his field studies. On the melanopic EDI scale, light levels only reached between 30 and 60 lux.

In its same report, the DOE found that striving for circadian metrics can impede meeting energy codes. Companies could curb would-be energy sinks by being choosy about where and when they use tunable lighting—perhaps skipping hallways and not leaving the lights on for the entire workday. Motion-detecting systems could help. A workplace could even designate one communal room as a "light oasis," where employees are encouraged to visit early in the day and whenever they need a boost. Morning meetings could be held inside. Vulcan has one of these rooms. And keep in mind that part of the point of circadian lighting is to reduce the use of light at the end of the day. Dimming or flipping lights off at night can save more precious energy.

Whether it is day or night, we could also get better at putting light "only where it is useful," according to the DOE. There is a lot of room for improvement: less than 1 percent of the light generated by indoor lighting ever reaches an occupant's eye. The solutions may be similar to what DarkSky and other advocates recommend for reducing light pollution. Circadian lighting could also prove more practical and conserve more energy if integrated with designs that, for example, automatically adjust artificial light levels with changing contributions of natural light from windows and skylights. The lighting could even be combined with ultraviolet disinfection and other technologies that have gained popularity since the COVID-19 pandemic.

As we know, daylight and fresh air act as disinfectants, too. Post-pandemic designers could heed Florence Nightingale's advice and incorporate more operable windows, creating buildings that are both more circadian friendly and COVID safe. In all this technological excitement, let's not overlook the gold standard: the sun.

THE 1845 EDITORIAL IN *THE LANCET* THAT OPPOSED THE WINDOW tax presented another argument in its attempt to enlighten readers: "For health, we cannot have too much light, and, consequently, too many or too large windows." The author also noted the different needs across different regions. London, the publication's home, sits well north of the Mediterranean. "In a climate like this, where the sun has so little power during by far the greater part of the year," the op-ed reads, "we should do everything in our power to admit as much light as possible into our houses, that we may bathe in it, and imbibe it by every pore." It goes on to caution against a pervasive bias at the time toward southern European building design, which was stingy with windows.

Windows and skylights can provide unrivaled natural light, fresh air, and dynamic views without using a watt of energy. Studies show people sleep better, score higher on cognitive tests, and get discharged from hospitals sooner when they receive natural light through windows. Yet we're still far from doing everything in our power to imbibe those photons. The rise of cheap, efficient LEDs appears to be a barrier. In August 2023, after years of debate, the University of California, Santa Barbara, ultimately scrapped plans for a radical new dormitory. It would have been the largest in the world. But equally, if not more, noteworthy was its daylight-free design: Almost all living spaces in Munger Hall would have lacked access to natural light. Panels of LED lights would have stood in for windows. Windowless rooms are already housing students at colleges, including the University of Texas at Austin, as more companies come up with versions of an LED "sun." It is incredible innovation. Yet experts raise worries about the quality and quantity of the light, not to mention the lack of a view. "While it may seem like a helpful supplement in a poor situation, it can also serve as an enabling technology, allowing building owners an easy excuse for not providing more access to daylight and views," said Lisa Heschong, the daylight expert and view proponent in Santa Cruz. "As in, 'Don't worry about working in the basement, we've put in the latest in circadian lighting.'"

Engineers, architects, and designers have come up with a few brighter, or at least more natural, ideas than Munger's faux windows. The Norwegian valley town of Rjukan has placed giant computer-controlled mirrors atop a steep mountain wall that flanks the town's center. The mirrors follow the sun and reflect its rays down on the residents, who would otherwise be living in shadows for almost half the year. Since 1928, a gondola has also transported people high enough above town for doses of winter sunshine. At smaller scales, windows, skylights, and mirrors can coax more daylight into buildings. That's also the idea behind the Circadian Curtain Wall. The design, which no one has yet built, includes a perimeter of windows that bubble outward, almost resembling flower petals from an aerial view. This floor-to-ceiling glass would provide occupants with wide-angle views of the outdoors while bringing natural light deep into the building. It would also reduce the heat and glare typical of rectangular glass buildings, because wherever the sun is on its path across the sky, the bubbled windows lessen the surface area exposed to direct sunlight. What's more, the curved glass would strengthen the building's skin.

Simple interior design considerations can go a long way in maximizing both natural and artificial light, while minimizing glare. A study of row houses in Boston found that minor alterations, like painting the walls white or utilizing areas closer to windows, could improve daylight exposure and, therefore, occupants' circadian rhythms. The authors also recommended broader changes, such as "discouraging use of the basement floor as a dwelling." In an office setting, bigger windows and low desk partitions can pull light deeper into buildings. Certain finishes on surfaces can reflect more light. Just the orientation of desks can go far. Feng shui, an ancient Chinese art of arranging living spaces and orienting significant sites and buildings to create harmony and balance with the natural world, instructs us to never have our back to the door. Circadian science urges us not to have our back to the window. Unfortunately, in many offices, the door is opposite the window. Perhaps we can strike a balance.

Our priority, according to Heschong, should be enhancing circadian light naturally through urban and building design. Then we can put "little technological fixes on top of that," she said. Such fixes could include smart windows, like those being used in Oregon State's trailers, which would help us lean even more on natural light. But she and other daylight advocates also acknowledge that artificial light may be a necessary supplement for our indoor lives—especially if you happen to be a hockey player, cafeteria worker, submariner, or sit in an office with your back to the window. As the gap between real and faux daylight narrows, novel technologies can fill in where natural light can't reach.

ABOUT A DOZEN MILES FROM VULCAN, AT HAZEN HIGH SCHOOL IN Renton, Washington, Thomas Walker teaches health science. His day starts far too early for his liking—the morning bell rings at 7:20 a.m. It is particularly tough during the winter when he arrives at school in the dark. But since the introduction of circadian lights at Hazen a few years ago, he told me he's noticed a night-and-day difference for himself and his students.

Walker also shared frustrations that have become another common refrain in conversations about circadian lighting. While classrooms throughout Hazen High and in at least two other district schools have the systems, few teachers use them. He listed various issues but pointed to the two most common culprits: a lack of training for the staff on how and why to use the lights, and a lack of power-safe memory for the lighting system control box. Because it takes in-class time to reset changes, he said, "most teachers give up the benefit in light of the inconvenience." Many may not even be aware of the benefits lost. Meanwhile, Walker's school district has not yet followed Seattle's lead on delaying the morning bell, which may be more potent than greater daytime light exposure for improving students' sleep.

Lighting experts and advocates told me that circadian lighting will only fulfill its promise once people are educated and systems are affordable, simple to install, and easy to maintain. Even then, technology notoriously goes awry, as *Saturday Night Live* played up in January 2023: "Due to a computer error, a school in Massachusetts has been unable to turn off its lights for over a year and a half. The students are doing fine, but the classroom hamster has gone insane."

One other prerequisite: the light needs to look good. Robert Soler, vice president of research for California-based BIOS, told me he wanted to create a lamp that was not "a flashlight in your face," but could still wake up the circadian system. In designing his company's SkyView line of lights, he kept the brightness levels moderate and concentrated the sky-blue wavelengths. This recipe, Soler told me, provides about three times as much biological potency as a standard LED. Still, you wouldn't necessarily know it based on the intensity or visible color. Soler included ample violet wavelengths, which are also usually lacking in modern indoor environments. And he incorporated longer wavelengths of light, too, making the combined color more palatable—or "more cuddly," as Naomi Miller, of the Pacific Northwest National Laboratory, put it. Light lacking long red wavelengths can make skin tones, fabrics, and other room finishes look bad.

As we've learned, adding wavelengths that don't fall near green on the spectrum is a hard sell. Current lumen-based lighting standards give little to no credit for light produced on the violet-blue or red ends, so the calculated energy efficiency takes a major hit. Red wavelengths don't improve melanopic EDI numbers either. "But I feel better about myself when I look in the mirror," said Miller. Plus, we're only now learning about red light's health benefits.

Little tweaks in a lighting recipe can pay big dividends for our clocks, while minimizing increases in energy use and unsightly photons. Here, we need to consider one more lighting metric: correlated color temperature. It is measured in Kelvins, commonly with values between 2200 to 6500. And it has nothing to do with whether a light is warm or cool to the touch. In fact, the terminology is a bit counterintuitive. Lower color temperatures appear as warmer hues, like the coruscating colors of a campfire, while higher color temperatures look more like a cool blue midday sky. Traditional incandescent bulbs emit a 2700-Kelvin light; a 5000-Kelvin bulb is typically labeled as "daylight." Different permutations and combinations of wavelengths, Soler found, can boost melanopsin-stimulating light during the day while barely changing how the photons appear to the eye. A mix that raises the correlated color temperature slightly from a standard "natural white" of 3500 Kelvin to

4000 Kelvin, for example, could increase the potency of that daytime signal to the circadian system by 58 percent, without needing to go as far as the stark white of 5000 Kelvin. Meanwhile, another special blend of photons could dial down the color temperature at night from 2700 Kelvin to the "soft white" of 2200 Kelvin, a color closer to candlelight, while dropping the circadian potency by 60 percent. These small changes combined, based on Soler's calculations, would ratchet up that critical day-night ratio nearly fourfold. He enlisted this strategy in designing the tunable SkyView table lamp.

In one study of the real-world effects of a melanopsin-stimulating lamp during the day, researchers had young adults sleep seven hours a night for more than a week before coming into the lab. Now slightly sleep deprived, "like most of us," said Harvard's Rahman, an author on the study, the volunteers were put through simulated workdays with a battery of tests while under one of two lighting conditions—either only ambient fluorescent lighting with a light intensity typical for offices of about 30 melanopic EDI lux, or with a supplemental blue-enriched task lamp that brought the light level up to a melanopic EDI of 250 lux, that recommended minimum threshold.

The gain in daytime alertness and cognition for those who got the supplemental lighting was remarkable—on par, Rahman said, with an hour or two more sleep nightly. And improvements in reaction time were equivalent to stopping about five football fields sooner when driving at highway speeds of sixty-five miles per hour. Rahman suggested that many of us accept our sleepy state as normal. We become habituated and hardly notice yet would probably test as though we were "inebriated," he said. And, no, these results do not suggest that we can skimp on sleep if we use a fancy light. Photons can never make up for all the losses associated with sleep deprivation.

Optimizing circadian lighting is getting easier with sophisticated simulation tools. Models can now evaluate daylight's direct illumination and its reflections within a building, based on realistic local climate conditions. They can also identify areas where daylight struggles to reach—allowing a designer to explore solutions through orientations, materials, windows, and electric lighting. The lighting specialist Marty Brennan has worked with a team at ZGF Architects and the University of Washington on open-source simulation software, called Lark, that can even predict how all of that light will affect occupants' inner clocks. In addition to standard measures of light, Lark incorporates metrics that give extra weight to wavelengths that stimulate melanopsin and neuropsin. Their algorithm can also help designers account for other considerations, noted Brennan. In a hospital, for example, you wouldn't want to pull out all the blue wavelengths if lighting is needed for an examination. Otherwise, doctors and nurses wouldn't be able to see a patient's veins.

The stakes are high for getting circadian lighting right. And the stakes are arguably highest for lighting inside hospitals and nursing homes which, unlike offices and schools, are occupied 24-7. Plus, the occupants are some of the most vulnerable to the impacts of circadian disruption.

A HERD OF US—PEDIATRICIANS, SLEEP DOCTORS, ARCHITECTS, AND me—shuffled through Cincinnati Children's corridors, passing walls painted just about every happy color, from Granny Smith green to pumpkin orange. It was September 2021, and we were going to check out the hospital's newly constructed neonatal intensive care unit, or NICU. There, inside patient rooms, a novel experiment was set to begin. The results could radically revise our advice to pregnant women and how we care for newborns—especially those born prematurely. Once again, it all comes down to the right light at the right time.

When a baby is born prematurely, they are prematurely disconnected from crucial clock messages from the mother. "We've made what I think are rather unintelligent assumptions about what we should do about that," said James Greenberg, the hospital's director of neonatology. Among those misconceptions has been that the womb is calm and dark and, therefore, that the NICU should be, too. Our tour stopped at the hospital's old, then still-operating NICU. The nurses were caring for a sick patient, so we did not walk inside. But we did peek through the open door. John Hogenesch, the circadian scientist at the hospital, was with us and took a light reading through the door: 10 lux. That dim lighting is very typical for a NICU, noted Greenberg.

Photons may begin exerting their influence at the earliest stages of life. As we've learned, light information passes indirectly from the mother to the developing fetus in much the same way the SCN relays the time of day to peripheral clocks throughout the body. Light may also filter directly into the womb. These cues are supplemented by the arrival of nutrients when the mother eats, circulating hormone levels, and a fluctuating body temperature. Greenberg also noted the ever-changing symphony of sounds, from blood pumping to bowels churning. By the second trimester, he said, the fetus is acquiring circadian information and developing circadian rhythmicity "of some sort." But what happens to fetal development when you take away those circadian prompts? This all sparked an idea: while doctors couldn't put a premature baby back inside the womb, they could at least try to bring a primary circadian cue—the light-dark cycle—back to the baby. They would fit the new NICU with tunable lighting.

Greenberg and his team hope that the new full-spectrum lighting system will improve outcomes for preterm babies and strengthen circadian rhythms of the care staff and parents. As I write this, they are closely watching the NICU's first patients to determine whether their instincts were correct. Studies already suggest that premature infants who get daily cycles of light and dark gain weight faster and leave the hospital earlier compared with those kept under near constant light or dark. "That's a metric everyone cares about because it's about $10,000 a day," said Hogenesch.

While getting cut off from the mother's circadian cues is only one of many challenges facing preterm infants, it could contribute to both short-term and long-term problems, such as learning disabilities, common in this population of kids. Richard Lang, the Cincinnati Children's researcher with expertise in nonvisual photoreception, worked closely with Greenberg on the project. He suggested that, as research further decodes the impacts of light on the growth of the brain and body, there are added reasons to be optimistic about the novel lighting. During development, ipRGCs begin responding to light before rods and cones do. And those early light responses could be crucial. One study in mice found that light stimulation of ipRGCs regulates the release of oxytocin in the brain to kickstart the creation of synapses and learning. As we walked the corridors toward the NICU, Lang further pointed out that some research links the development of autism, schizophrenia, and many other diseases to the season of birth. Scientists have seen similar patterns for immune function. Such differences across seasons could reflect differences in early life light exposure.

So far, the findings only point to a correlation and are not proof of seasonal timing as a cause of any conditions. But it is an intriguing connection. "We're gathering evidence to suggest that light pathways regulate development, in particular neurodevelopment," said Lang. Blue light and violet light could be key ingredients for healthy eyes and brain. Levels of these wavelengths vary across the year and with the overall length of the day. And, again, many of these wavelengths are all but absent in today's indoor environments. The team has integrated them into the NICU's new lighting system.

Our first stop in the NICU was room 52. Robin's-egg blue accents and white wallpaper with a delicate leaf pattern covered the walls. A rectangular tunable LED fixture was mounted on one of them, pointed upward. A white ceiling reflected and diffused the light. Brennan, also collaborating on the project, carried his high-tech spectrometer on the tour. Needless to say, it was picking up a lot more light than it had in that Seattle market. This time, as he eagerly showed me, the device informed us that were basking in about 1,400 lux, with peaks in the violet, blue, and red parts of the wavelength spectrum.

A LITTLE MORE THAN A YEAR AFTER MY VISIT, A NEW NICU ROOM became a temporary home for Madison Whipp. She was born about ten weeks preterm, with complications, including Down syndrome, a malformation in part of her gastrointestinal tract, and heart and lung issues. Doctors gave her less than a 1 percent chance of survival. At two days old, Madison was flown from her hometown in West Virginia to Cincinnati Children's for surgery. She would ultimately stay at the hospital for five months. Every day, the lights in her room brightened in the morning, peaked through midday, and softened in the evenings. Madison was discharged on December 23, 2022, just in time for Christmas and months before her parents expected to bring her home. The feat earned her the nickname "Miracle Whipp."

I spoke with Kylen Whipp, Madison's dad, in late January 2023, just after Madison turned six months old. "I don't know if it was because of the lights, but she has gotten into a really good rhythm during the day," he told me. "She sleeps totally through the night now." In fact, he added, she was "getting into that rhythm" before she even left the hospital. The family will never know if the lighting influenced Madison's survival, Christmas homecoming, or regular rhythms, but the science increasingly supports the strategy's benefits.

Long before the NICU lighting project was conceived, Lang routinely advised pregnant woman to spend time outside in order to bathe herself and her baby in full-spectrum sunlight. By keeping her own rhythms entrained, the woman might also lower her baby's risk of ending up in the NICU. Lang and other experts emphasized the continued importance of consistent circadian cues for the developing infant, including through breast milk. Really early preterm babies in the NICU usually don't get milk. Then, as they slowly recover, milk is introduced.

But that milk is typically stored and often pooled, once again washing out critical time-of-day hormonal cues from the mother. As we've learned, our circadian system evolved with some helpful redundancy. While it may be challenging in the modern world to get all the zeitgebers just right, we might improve a baby's chances of success by dialing in at least one or two.

To inform the appropriate daily mix of light—wavelengths and intensities—that "reach babies' eyes in their bassinets," said Brennan, he encouraged Lang's team to collect light data from the hospital's rooftop. They obliged. I got to peek up at the small glass sphere mounted atop one of the research buildings, the highest point on campus. Every minute of every day, it collects the spectrum and intensity of light from the skies over Cincinnati—and connects by a two-hundred-foot fiber optic cable to a spectrometer inside the building. This black ball is much smaller than the red one atop the observatory in Greenwich. Still, it has a similarly impressive 360-degree view—from the tops of the city's skyscrapers and the Kentucky border to the south, to the elephant house of the zoo to the northwest.

What the sensor collects here differs from what it might gather from the skies over Anchorage or Dubai. "Daylight is the gold standard. But daylight where?" Brennan said, noting the natural influences of latitude, climate, and geography. In recognition of this variation, Brennan and the University of Washington have begun compiling a database of natural light from various regions of the world. I joined him a few months later atop a roof at the University of Washington in Seattle. Researchers are doing the same in other locations, including Colorado and Spain. The goal is to develop indoor lighting recipes for schools, businesses, hospitals, and other buildings that re-create a semblance of local environments—to a point. "We don't necessarily want to mimic the lack of daylight at extreme latitudes or the incandescent spectrums when the sky is full of smoke particulate from fires," Brennan told me. If I could choose, I would certainly opt for the local light environment of a place like Sacramento over Seattle.

MICKEY AND MINNIE MOUSE SMILED AND SWAYED ON A WINDOWSILL—the bobblehead dolls' dance powered by rays of California sunshine. A dozen feet away, a woman sat in her wheelchair, engrossed by a soap opera. She was back in her room after winning a round of bingo and would soon roll down the hall to lunch at the Bistro café. The stretch of hallway outside her room is known in this Sacramento nursing home as Bamboo Lane. Since the ACC Care Center replaced Bamboo Lane's fluorescent lights—which stayed on full blast day and night—with tunable LEDs, she and other residents reportedly slept better and experienced better moods. Melanie Segar, an administrator at the center, described one resident as having gone from being angry much of the time to being in a "happy place." "That's the best we can hope for with dementia," she told me.

Light is a critical consideration from our first to our final days of life. By age eighty-five, a person needs roughly seven times the illuminance of a twenty-five-year-old to stimulate the circadian system. Blue is the part of the light spectrum most filtered by the eye's yellowing lens and cataracts. A narrowing pupil and more time spent indoors also conspire, along with aging's other detrimental effects on the brain, to create a vicious spiral of circadian clock mismatches, reduced rhythm amplitudes, erratic sleep patterns, and cognitive decline. Research points to strategies to slow that cycle. One study linked cataract removal with a 30 percent reduction in the risk of dementia. Many more show benefits of the right light at the right time. Increasing the daily light-dark contrast, according to research by Figueiro and others, shows promise for improving mood and sleep for people with Alzheimer's and Parkinson's diseases.

When Scott Stringer started as the ACC Care Center's medical director in 2014, he began giving residents with behavior and depression issues an unusual prescription: a couple of hours of morning sunshine.

That's generally a fillable order in a city boasting two hundred sixty-five annual days with sun. But getting everyone a daily dose of courtyard rays is not practical for a ninety-nine-bed facility. And this building did not have the same massive windows as the nursing home I visited in Denmark. Thankfully, the introduction of circadian lighting at the center meant Stringer's light prescriptions were no longer limited to those patients he could roll outside.

Manipulating artificial light may be a valuable non-pharmacological tool to help regulate the circadian rhythms of older adults, especially those who have dementia. If tunable LEDs can compensate for some of the inevitable circadian losses, the technology could raise residents' well-being and cut health care costs, Stringer told me.

When I visited in 2017, the ACC Care Center had recently installed its first tunable LEDs in a couple of residents' rooms and the adjacent corridor of Cherry Lane, as well as a few other areas. In addition to energy savings of 68 percent relative to the old fluorescent system in the corridor, a preliminary US Department of Energy study found that the three residents living in the two rooms had significantly fewer incidents of yelling, agitation, and crying following the installation compared with the preceding three months. The nursing staff also noticed that these residents began consistently sleeping through the night. "I didn't think much about lighting before. But I've become a true believer after seeing how it's affected residents," said Segar. Once Cherry Lane got the tunable LEDs, residents from other corridors began hanging out there. "And their family members began asking when they would be getting the lights, too," said Segar.

After the center replaced the lights in all halls, common areas, resident rooms, and bathrooms with tunable LEDs, as part of a broader remodel of the facility, the DOE conducted a follow-up study. Researchers randomly assigned corridors to lighting that transitioned in intensity and color across the day or to static lighting that mimicked the old fluorescent lighting. On average, residents experienced half as many nighttime sleep disturbances with the tunable compared with the static lighting.

Surveys suggest that many more nursing homes worldwide could benefit from changing a few light bulbs. One study of fifteen nursing homes in Norway found a "ubiquity" of insufficient melanopsin-stimulating lighting in dementia units. In another study comparing outcomes in two pairs of nursing homes in Wisconsin, Steve Lockley and his coauthors found a 43 percent reduction in falls among residents in homes equipped with tunable LED lighting compared with residents in homes with standard lighting. Because they saw a larger reduction in falls overnight when lights were dim compared with during the day, it was unlikely that a change in visual acuity explained the improvement.

While tunable systems are nice, Lockley suggested that simpler and less expensive lighting solutions can suffice in many contexts. He told me about an intervention in Australia for patients with acquired brain injury, another population prone to sleep problems. The research team went into the patients' homes and changed out light bulbs. They installed higher-wattage bluer ones in the ceiling and lower-wattage redder ones in bedside and table lamps, all purchased at a local Bunnings hardware store. They instructed the patients to use ceiling lights during the day and lamps at night. After eight weeks, the sleep of patients with the new lighting significantly improved. "It doesn't have to be an automated system," Lockley said. "Ultimately the eye doesn't care. It's the photons the eyes detect that matter."

11

Hacking Rhythms

Despite all my preaching to friends about how to care for the amazing little clocks inside their bodies, I still regularly went sleepless in Seattle. And, more generally, I struggled to stay on my A game. Was I even practicing good circadian hygiene myself? A full audit was in order.

It was a typical overcast February day in Seattle when I began with an assessment of my light intake. How did my days and nights fare in light of the recommendations of at least a melanopic EDI of 250 lux during the day, below 10 lux in the evening, and below 1 lux while sleeping? I'm fortunate to have a southeast-facing picture window by my desk. And I'd learned that the window's glass lacks a lot of the typical energy-efficiency coatings impermeable to healthy wavelengths. As a bonus, no buildings obstruct my daylight or my view, just a few big cedar and maple trees across the street that are frequented by flitting finches and cawing crows. I also have French doors that lead out to a small balcony and another window in my kitchen. Daylight streams into my little apartment. I figured this all meant that my eyes stayed satiated with photons. I wasn't wrong. But I wasn't quite right either.

The sky was a reasonably bright light gray that morning. Erik Page, CEO of Blue Iris Labs, had recently sent me a prototype of his research-grade spectrometer. The Speck is shaped into a rounded triangle, about an inch wide and made of black plastic with a white spot in the middle—more conspicuous than the cute LYS button, but considerably more powerful, too. Through a web portal, I could track the spectrum, intensity, and even the calculated melanopic EDI of the photons it collects in near real time, with graphs in a rainbow of colors. Page reminded me that while our body logs this data all the time through our visual and nonvisual systems, we generally don't consciously notice. The Speck should help me notice.

I threaded a ribbon through the notch at the top of the device and looped it around my neck. Now, alongside my LYS clipped on my shirt, it could take regular readings of my light environment. I was supplementing the day's subpar natural light with artificial light from a floor lamp equipped with adjustable Philips Hue bulbs, dialed up to their bluest and brightest, and another small lamp on my desk. I'd set that to its full brightness as well. I also flipped on the overhead kitchen and dining room lights. My home looked luminous to my eyes. Yet, standing in the middle of the living room, in sight and within about a dozen feet of three windows and lots of light bulbs, the Speck only registered a melanopic EDI of 75 lux. My LYS app concurred: "This light is making you DROWSY," it read.

Okay, fine. That wasn't the spot where I spend most of my time anyway. I walked toward my office, which occupies the front left corner of my living room by the big window. That's where I sit or stand during most daylight hours. I stopped on the way, about a foot in front of the middle of the window. Bingo. The Speck reading shot up to a melanopic EDI of nearly 11,000 lux. I took a small step back, to the distance I sit from the window while at my desk: about 4,500 lux. Then I turned ninety degrees to the left, toward my computer screen. The light reaching the Speck dropped to approximately 800 lux. Finally, I took a step forward, paralleling the window, and sat down at my computer. My reading plummeted below 150 lux. That couldn't be right, I thought. I took repeated readings, retracing my steps to double-check those numbers. Indeed, it seemed a twist and shuffle step were enough to nearly blind my ipRGCs.

Shocked and a bit frustrated, I decided to clear my head. I also had some making up to do for my ipRGCs. I bundled up for the cold and headed outside with the Speck still around my neck. The sun punched through the gray a couple of times, warming my face and the Speck as I walked in the park across the street. The data nerd in me was eager to see what it saw. Back at my desk a short while later, I opened the web portal: The partial sunbreak had registered a melanopic EDI of more than 29,000 lux—and light levels for the entire walk proved substantially higher than any of my indoor readings. But now, from where I sat at the computer, my light levels hovered back down around 100 lux.

I kept up the close light monitoring for a few days. A window on my computer continuously displayed my data feed. At one point, I amused myself and a friend by wearing the Speck suspended on my forehead as a third eye—a more accurate reflection of what my third photoreceptors saw—while we walked the same park trail. I didn't get nearly as many funny looks as I'd anticipated. I also retested my light exposure another afternoon when the clouds parted to reveal a clear blue sky. Outside, the Speck recorded a melanopic EDI near 80,000 lux. But sitting at my desk, my light exposure remained in the low two hundreds, still below that recommended threshold. Okay, what gives? I decided to try an experiment.

When photons of light pass through a window and travel into a room, they quickly spread out. It doesn't take much distance to lose brightness, although light-colored or mirrored walls, ceilings, floors, and furnishings will reflect and amplify light levels. My desk is only a couple of feet from a window, but I realized little of its glass was actually in front of me—or, more relevantly, in front of my eyes. When I pulled my desk only about a foot away from the left-side wall so that more inches of window stretched in front of me, the Speck's readings shot up. They now hovered between 400 and 500 lux, despite the sun now having gone behind light clouds. Whenever I turned to gaze out the window, the readings jumped into the four digits. I remembered the tip from Lisa Heschong that daylight striking our eyes as we look directly out a window can be surprisingly more potent than daylight that reaches them when we gaze elsewhere within an indoor space. Now, I could see a very bright side to my regular distractions, like the other day when I paused to watch a bald eagle and crow swoop and squabble atop a tall cedar. And a few days before that, when I saw a young unicyclist ride down the trail while juggling. Definitely a humbling moment, and clarity as to why my "third eye" drew little attention. Anyway, I now have a name for these intermittent peaks of daylight and view: Heschong calls them "circadian snacks."

My little rearrangement wasn't foolproof. On one of those "Big Dark" days, as we call them in Seattle, with heavy gray clouds turning day almost into night, light numbers at my desk plunged back into the double digits. Regardless of the weather, I noticed that my light readings regularly dropped later in the day as the sun journeyed across the sky and beyond my window's direct view. Even on the sunniest days in the winter, it fell below a melanopic EDI of 250 lux by midafternoon. To catch more photons at my desk, at least during the more crucial morning exposure period, I upped the brightness on my laptop and second monitor and added the beams of a compact light therapy lamp. Even with the darkest skies, I was back in the 400s. I also tested out the more palatable BIOS SkyView table lamp, with its shades of "sky blue" emanating through handblown glass. At the daylight setting, a warmer glow at the bottom melds into a bluer light at the top. Sitting in front of that lamp alone got my numbers well above 500 lux. I now had multiple remedies to escape the dark, and ideally abate some of the gloominess I frequently feel through the winter.

Seeing these numbers motivated me to get outside more, too. As I sat on my balcony one afternoon, a sunbreak registered a melanopic EDI of around 70,000 lux. I tried holding my purple sunglasses in front of the Speck sensor. The reading dropped to 5,000 lux—still a powerful dose. I got comparable numbers under the shade of a hat. I later repeated the tests on a cloudy day. Even when skies aren't clear, as I'd learned, nature's circadian light is almost always potent. My Speck registered around

1,600 lux. But it dropped to just under 200 lux with sunglasses. It fell further, to just above 100 lux with my blue-light-blocking glasses. The data decided it: I would continue wearing sunglasses on bright days to avoid squinting but heed the advice of the experts and skip the glasses for at least some of my time outside, especially during morning hours. And I would never wear my blue-light blockers during the day.

Many people, often with encouragement from optometrists, believe they should wear blue-light-blocking lenses during the day to protect their eyes, especially while at their computer. Andrea Wilkerson, a lighting research engineer at the Pacific Northwest National Laboratory, was annoyed: "It's such a marketing thing for eye doctors." The American Academy of Ophthalmology states that the amount of blue light coming from screens has never been shown to harm or strain our eyes. Wearing these glasses all day, however, could blunt the valuable melanopsin-stimulating light reaching our ipRGCs. I pledged to hoard as much light during the day as possible, in part to build a protective buffer from detrimental photons at night. Of course, that would be the time to don those glasses.

IT'S A LITTLE AFTER 9:00 P.M. AS I TYPE THIS. ALTHOUGH THE REST OF my apartment is dim—a melanopic EDI of less than 1 lux—the Speck is registering 71 in front of my computer. After I dial back the brightness and turn on the Windows Night Light feature on my laptop, which mutes and warms the color temperature of the screen, it drops to 13 lux. That is still above the recommended nighttime max of 10 lux. I finally duck under that benchmark, at 4 lux, when I hold the blue-light-blocking glasses between the Speck and screen. I find very similar numbers following the same sequence for my iPhone.

All blue-light-blocking glasses are not made equal. One clear indicator is the clarity of the lenses. For the glasses to block a substantial amount of blue light, they generally need a yellow-orange tint—even if that might make them less socially acceptable to wear. I'll admit that I'm not thrilled when my glasses turn the pretty blues in my world to gray, reminiscent of those bland bunker blueberries and typewriter. Fortunately, such annoyances might be avoidable in the future. In addition to fiddling with the wavelengths emitted by lamps, engineers are also busy designing screens, glasses, and other tools that filter or enhance certain wavelengths without altering how the light looks to us.

Concerns about the blue light emitted by screens had yet to reach high on the social radar when Lorna and Michael Herf developed f.lux, the first software to adjust the color temperature of a computer's display. It predated the Night Light. The couple lived in Los Angeles at the time, and Lorna frequently painted in their loft. They had retrofitted the space with daylight-mimicking, full-spectrum lights directed at her canvas. This kept the colors of her paint coherent day and night—opposite of the contrasts desired by our circadian rhythms.

One night in 2008, Lorna walked downstairs from the loft to where her husband sat in front of the fireplace. Complementary warm evening colors filtered in through the windows and from incandescent lights. Michael, then a software engineer at Google, was working on his laptop. Lorna was struck by that laptop's intensity and cool light, which more closely matched the lighting in her loft. She told her husband that his screen color looked off: "Your laptop is still daytime." Michael laughed. "I can fix that," he said.

The couple created the first version of f.lux after Michael left Google in 2008. "It began as an art project," said Lorna. "And it definitely took a hard turn into the sciences when we realized, 'Oh my God, look at all these people who are holding these devices ten inches, six inches from their nose, especially kids,' right?" We've all been awash in light without realizing it, the couple told me.

Reducing short-wavelength light from displays can lessen the melanopsin-exciting action of the light. But most studies conclude that the effect is moderate. Again, while melanopsin is most sensitive to wavelengths around 480 nanometers, it will still respond to other wavelengths. The further those wavelengths fall from 480 nanometers, the brighter the light needed to achieve the same stimulation. And we still don't fully understand the circadian contributions of our cones, which peak at different wavelengths—and could be extra stimulating if the light contains contrasts or fluctuations, as it would with a TikTok video or streaming movie. Simply dimming your screen can be an easy and effective way to tell your body it's nighttime. In another Speck test, I found that setting my iPhone's brightness to its lowest level—still plenty bright to read in my dark bedroom—registered less than 1 lux, even without the Apple Night Shift feature turned on. Plus, this way, I could still see the true color of clothes during any late-night online shopping. Yes, I'm aware this also reflects poorly on my circadian hygiene.

Both Lorna and Michael recognized that using a program like f.lux is not enough to ensure a distinction between day and night for our circadian system. We need to consider all the photons we immerse ourselves in throughout the day. "If your day is one hundred times brighter than your night, you might be doing pretty well from a circadian perspective. But if your day is two times brighter than your night, you might not be doing very well," said Michael. With the help of software-controlled lighting, he added, the Herfs' house "does about fifty-to-one from day to night." Michael recalled a visit from friends who claimed to be night owls, regularly staying awake into the early morning hours. "They were asleep on our couch at nine thirty," he said.

Experts tipped me off to a few other strategies for shedding photons at night. I have now arranged eighteen flickering electric candles around my apartment. A remote control ignites their faux flame. The only other lights I leave on are a table lamp with a three-way bulb set to its weakest and warmest and undercounter kitchen lights. It's cozy or, as my new friends in Copenhagen would say, "hygge." After discovering that my bathroom lights shine a melanopic EDI of 20 lux at my eyes when I stand in front of the mirror, even with only two of the three bulbs lit because I've been lazy about replacing the third, I now avoid flipping that switch on at night. Instead, I've plugged an amber night-light into the outlet shared with my Sonicare. It provides plenty of light to brush my teeth and for any late-night bathroom trips. And, I must say, I appreciated my ex-boyfriend's red toilet light. I find all these edits more pleasant than wearing blue-light-blocking glasses. I reserve those for when I'm working at my computer late into the night, which I hope to do less often once I finish writing this book. I am also now listening to audiobooks rather than enlisting light to read in bed.

The Speck confirmed that my bedroom sleeping environment stays below 1 lux, at least until the sun comes up. That's good enough for the winter months. But when the sun rises before I do, from mid-spring into the fall, I knew its rays would again sneak through and around my drapes and beam more than a couple of lux at my eyes. So I upgraded to thicker blackout curtains and bought my first eye mask, the comfy kind with contours so as not to tickle my lashes.

Meanwhile, it was time to extend my personal auditing. Other factors influence my circadian clocks every day—from when I eat, to my caffeine and alcohol intake, to my sleep schedule. "You can't just focus on one aspect," said Pacific Northwest National Laboratory's Andrea Wilkerson. She referred me to an article in *The Onion* that, she said, highlighted "the insanity of all of this." The headline read WOMAN WHO DRINKS 6 CUPS OF COFFEE PER DAY TRYING TO CUT DOWN ON BLUE LIGHT AT BEDTIME. Wilkerson had a point. Multiple suspects were probably implicated in my circadian troubles.

MY NEXT STEP WAS AN HONEST ASSESSMENT OF MY EATING AND drinking patterns—logging what I put into my body, and when.

For most days of the week, I calculated a thirteen- or fourteen-hour window between my first cup of coffee with oat milk and my last sip or bite of calories in the evening. That non-fasting window ticked closer to fifteen or sixteen hours when I had social plans. The timing of my daily meals didn't seem to be the main problem. My breakfast and dinner tended to span from around 9:00 a.m. to 8:00 p.m. It was the morning and evening beverages—frequently caffeinated or alcoholic—and the late-night snacking that got me. As I'd learned, every splash or morsel shrinks the overnight fast.

I started collecting advice from experts on how best to improve my diet, in a sustainable way. I wasn't ready to completely cut out coffee, wine, or a social life. The Salk Institute's Emily Manoogian claimed that a hot cup of water is "shockingly helpful" to delay her need for coffee in the morning. Still doubtful at the time, I conferred with other circadian scientists on that point and initially embraced advice that I found more agreeable: I could compromise by keeping my first cup of morning coffee black. But soon I began holding off as long as possible on both calories and caffeine to let my cortisol and adenosine finish their morning routines. I also curtailed my daily caffeine intake, including my beloved dark chocolate, to before noon and my cups of coffee to two, max.

I limited alcohol, too. Again, in line with the science, my Fitbit confirmed far superior sleep when I didn't drink. I vowed to swap that late-evening wine for herbal tea, at least on most days. If I did imbibe, I tried to cut myself off a few hours before bedtime.

I also tightened my eating window. "It's not about cutting meals. It's about scooching in," Manoogian had told me. "And it doesn't have to be extreme." Even an eleven-hour window could yield benefits. Now, most days of the week I consolidate my calories to between roughly 9:30 a.m. and 7:30 p.m. I felt like I was winning with ten hours. And my body adapted relatively quickly. Within a week, I no longer felt hungry before bed. I didn't wake up hungry either.

There were additional perks. Eating earlier meant I could take advantage of lunch and happy hour prices. I do worry how adhering to this plan will jibe with my social life long term, however. Thankfully, Manoogian told me that taking a cheat day here or there—just as Victor Zhang does with those pig roasts and wine nights—won't mess you up too much.

I began to enlist other circadian tricks gleaned from my conversations. I had occasionally taken supplemental melatonin to help me sleep. Apparently, I was far from alone. The use of melatonin supplements by adults in the US rose more than fivefold between 1999 and 2018. Experts are even more concerned about sharp increases in melatonin use among children. Global sales of melatonin supplements are projected to surpass 3.5 billion USD by 2030. Yet, as sleep experts told me, melatonin is "not a sleeping pill." In fact, it can worsen sleep and circadian problems, depending on how much a person takes and when. For the record, many sleeping pills on the market can cause greater harm than good, too.

Now, the rarer times I use a melatonin supplement for sleep, I stick with reputable brands. Melatonin content can vary significantly from what a label might say. I also follow the advice I heard from Cincinnati Children's John Hogenesch and others: take at most a couple of milligrams and do so two or three hours before bed—around when your melatonin levels should naturally ramp up for the night. Studies show that less than one milligram, which is the most the average adult naturally produces in one day, can be as effective as five or more milligrams in helping you fall asleep. And that strength carries lower risks of waking up groggy the next morning or messing up your circadian system's natural response to melatonin.

We know that melatonin supplementation can advance or delay the circadian clock depending on when it's taken, a power especially useful for shift workers and jet-lagged travelers. It can also help any of us with clocks that run slightly longer than twenty-four hours. Taking melatonin a few hours before bed—before your body would naturally start producing more of the hormone—nudges the circadian system to dial forward and promotes earlier sleep. However, and this is an important point, taking the same dose at bedtime can be counterproductive and cause a slight delay in your rhythm. This could make falling asleep that night, and the next night, more difficult. I now suspect I exacerbated my past multiday bouts of insomnia by attempting to take supplemental melatonin before bed as a remedy. In fact, I was in another rut when Hogenesch shared this tip. Sure enough, that night I took a small dose of melatonin at 8:00 p.m. and was sound asleep by 11:00 p.m. I had been wide awake until 1:00 a.m. the previous night.

I also learned that the ideal room temperature for sleep, at least for adults, is in the midsixties Fahrenheit. Directed in part by our circadian rhythm, our body's core sheds two or three degrees Fahrenheit at night by dilating blood vessels and increasing blood flow to our extremities. The surrounding environment needs to be relatively cool for this to work efficiently. Yet I was almost always living and sleeping with my thermostat set within a degree or two of seventy, day and night, a far cry from the natural fluxes outside. I began turning down the thermostat at night and enlisting a portable air conditioner during hot summer days.

Another practical and free home remedy: a warm bath or shower before bed can promote sleep by encouraging blood flow to the surface of the skin, allowing heat to radiate out. Because we most readily release heat from our hands and feet, warming these may further speed this transition to sleep. Mattress companies, including Sleep Number, are now marketing climate-controlled beds that aim to cool your core and warm your extremities simultaneously. While I haven't gone as far as purchasing one of those new mattresses, I am now heating my hands and feet before I crawl under the covers—which feels all the more necessary after lowering my thermostat. Unfortunately, with rising nighttime temperatures due to climate change, a cool sleeping environment will become less feasible for many people worldwide—especially those unable to afford a luxury mattress or air-conditioning.

Among the other circadian tips I gathered was to take naps, but limit their length and restrict them to early in the day. A snooze of just ten to twenty minutes can boost energy and alertness. A longer nap, however, may not only induce the paralyzing grog of sleep inertia but also reduce the homeostat's sleep pressure, making it more difficult to fall asleep that night. The later the nap, the worse the bedtime roadblock. I find that a short doze, especially if followed by a short walk under the daylit sky, helps power me through the midafternoon trough. I now regret not taking advantage of the available nap pods when I worked in the HuffPost newsroom.

Finally, the developing science inspired me to work on the regularity of when I take in that first and last photon and calorie, when I exercise, and when I go to bed each night. The circadian system craves consistency. In fact, a study published in January 2024, just as I was completing this book, found that the day-to-day stability of when we sleep is a stronger predictor of mortality than the duration of our sleep. Another study published that same month linked a more active wake period and a more regular, restful, and earlier sleep period with lower rates of cardiovascular disease and obesity.

I still struggle to keep a regular schedule. But I'm noticing that, when I do, I feel better and am more productive, happy, and motivated. Enough so that I am motivated to stick with it. That RISE app offers reinforcement. It estimates when my melatonin levels will begin to climb based on recent sleep and wake times, and gives me an ideal window for falling sleep, as well as the expected peaks and troughs of my energy across the day. It is nearly spot on. Whenever my eyelids grow heavy in the midafternoon and I check the app, I can see that I'm at the bottom of that valley. And as the graph dips again in the hours after that evening second wind, I notice a drop in my energy and alertness, and in my heart rate on my Fitbit.

I've made progress in waking up at the same time every day without an alarm clock. Of course, the switch to daylight saving time in March threw me off. But the persuasive social clock and the sun's progressively earlier risings eventually swayed me earlier. As I write this, I am waking up between 7:30 and 8:00 a.m. and going to bed by 10:30 p.m. or 11:00 p.m. The more consistently I abide by those times, the easier it is to naturally fall asleep and wake up at those same times. In between, I try to live my days respectful of my rhythms. I aim to get outside for at least a short walk by 9:00 a.m., typically a stroll to my local coffee shop, and then pair that coffee with breakfast. According to both RISE and my body, my energy peaks at around 10:00 a.m. and again at 6:00 p.m. When I need an extra boost in between, I opt for a short power nap or a hefty dose of photons—or both. Unless it's pouring rain, I now seek doses of daylight from my balcony year round—with wintertime support from a puffy coat, beanie, and hot cup of tea. Then, at least three hours before bed, I begin blocking the blue light, cutting the calories, abstaining from alcohol, and considering melatonin—only if necessary. As bedtime approaches, I turn down the thermostat.

In short, here are three core clock rules I now try to live by:

1. *Rise and shine*: Get twenty to thirty minutes of daylight, or the equivalent, first thing in the morning—ideally, after waking up without an alarm. Continue to maximize light throughout the day.
2. *Dim after dusk*: Minimize exposure to electronic screens and other artificial lights at night.
3. *Bite when light*: Aim to eat and drink calories only while the sun is up. The applicability depends on the latitude and time of year. As an alternative, if you go to bed in the vicinity of 11:00 p.m., like me, you can remember to *eat before 8*.

Most of my circadian interventions are fairly simple, some borderline boring. However, the hacks can get a lot more interesting.

"COFFEE?" KIM ADELAAR ASKED. I HAD JUST ARRIVED AT THE IJMOND police station in Beverwijk, North Holland. And as much as I knew better—it was after noon, my cutoff—I obliged. Travel had worn me down, so I granted myself a cheat. Adelaar handed me a cup from the coffee machine. Then I noticed she didn't make any for herself. "I don't drink coffee," she said. "The glasses are my drugs."

Adelaar referred to a pair of glasses that, at first glance, probably wouldn't turn heads—at least not any more than a pair of BluBlockers. The black plastic frames look normal enough; if anything, they are sleek and sporty. But press a small button by the temple, and the glasses turn space-age. Tiny LED lights embedded in the top of the frames beam blue light at an angle toward the eyes. A reflector on the bottom of the glasses amplifies the effect. With the lights off and orange lenses swapped in for the default light blue–tinted ones, the glasses can also do the opposite: block ambient blue light from reaching the eyes. Both uses, she said, saved her while working rotating shifts at the police department.

These particular glasses are sold by the company Propeaq. I was visiting the station with the company's CEO, Toine Schoutens. We were in the middle of a long day crisscrossing the country to meet with a sampling of his clients. Schoutens, also CEO of the circadian consulting company FluxPlus, once worked with Philips Lighting to develop smaller and smaller versions of therapy lamps and wake-up lights. In the 2010s, he took his shrinking game to the next level.

He told me that the concentrated blue light from a pair of Propeaqs negates the need to sit in front of a big box light. Rather, the lights can be worn for twenty to thirty minutes while going about your morning routine, with the dose remaining controlled and consistent. Propeaq glasses lean on decades of circadian and lighting research, and I heard positive anecdotes. But peer-reviewed studies of the glasses are a mixed bag. One quasi-experimental study of night shift nurses showed improvements in fatigue and well-being when they paired the glasses with naps, while another of night shift security guards failed to find a significant beneficial effect. In a survey of users with Parkinson's disease, the majority reported improvements in sleep, mood, and motor symptoms.

Propeaq isn't the only company filling this niche. Among other options on the market are the AYO blue light therapy "glasses." Its makers did away with lenses and opted for a band of blue LEDs that look straight out of *Star Trek*. The US Department of Defense reported improved sleep and performance for submariners who used the glasses for about forty minutes after waking up and wore another company's blue-light-blocking glasses for about two hours before going to sleep. Sarah Chabal, a research psychologist formerly with the Naval Submarine Medical Research Laboratory in Groton, Connecticut, led that research. Because of the round-the-clock shifts on the submarine, she told me, it is more practical to attach interventions to sailors' heads than to use one-light-fits-all fixtures. The crews appeared to embrace the tool, as evidenced by high compliance. I asked Kevin Brockman, the former submariner, about the glasses. Sailors aboard a different vessel had participated in the trial. “We heard about it, and made fun of them,” he told me.

YARA VAN KERKHOF, A DUTCH SHORT TRACK SPEED SKATER, IS ADelaar’s sister-in-law. Her team has worn Propeaq light glasses for years. Van Kerkhof said they proved very useful leading up to and during the 2022 Olympics in Beijing to adapt to the new time zone, align their peak performance times with scheduled race times, and give themselves a little lift before races. The blue light and the blue-light-blocking functions each had their time and place. Van Kerkhof and her team won a gold medal and set an Olympic record in Beijing. Just how much the glasses contributed to their success is impossible to know. Many factors were at play, of course. “It’s really nice to look for those things that can help you, even if it’s one percent,” she said. “In sports, everything counts.”

One summer, a couple of years before the Olympics, Adelaar borrowed the glasses from Van Kerkhof. The skater was in her offseason and not traveling. Adelaar’s body, meanwhile, was perpetually “traveling” between time zones while she worked rotating shifts as a police dispatcher. She hoped the glasses might help keep her clocks in the same time zone and her eyes open during long nights. Each afternoon before a night shift, she used the blue-light-blocking orange lenses to induce sleepiness so she could nap. When she woke up in the evening, she put the glasses back on with the blue lights activated. Then she went to work. “I tricked myself that it was morning light,” she said. She followed this routine for two years. For Adelaar, it was a welcome improvement from her previous eight years working shifts as a traditional police officer on the streets, struggling to stay awake and fall asleep at unnatural times. Adelaar’s rave reviews motivated Van Kerhof to use them even more. Soon, both had a pair.

Today, as the department’s vitality coach, Adelaar hopes to make light glasses standard issue with uniforms for every police officer. One of the officers at the station, wearing a navy blue uniform with fluorescent yellow stripes, modeled a pair of Propeaqs while sitting atop the hood of a white Mercedes-Benz marked with red and blue stripes and "Politie." The police car itself was, of course, topped with bright blue LED lights. "They are used to blue lights," Adelaar joked. A light bulb seemed to go off for her: "They could just turn the blue lights on for half an hour and look at it. We don't even need glasses."

While it was still early in the rollout, Adelaar reported promising feedback from officers. The anecdotal accounts from her colleagues with PTSD were impressive. "They don't have nightmares anymore," she said.

Because the Dutch police officers are constantly rotating their work hours, which is less conducive to realigning clocks, they found the blue light's direct alerting effect most useful. Manipulating inner clocks with strategically timed blue light was the greater attraction for the Dutch postal service workers I met at their logistics headquarters in Nieuwegein, about an hour south of Beverwijk, past several charming windmills and massive fields of solar panels.

While shades of blue permeated the police station, traditional Dutch orange was everywhere at PostNL—chairs, lamps, floors, uniforms. I sat in a plastic chair in a windowless meeting room as Schoutens presented to a group of workers, mostly truck drivers. Maybe it was the jet lag, or maybe the weak lighting and all the orange, but my eyes felt incredibly heavy. As Schoutens explained in Dutch the basics of circadian rhythms and how that science informed tactics the workers could use to optimize their clocks, a pair of Propeaq glasses with orange-tinted lenses went around the room for workers to try on. They perfectly matched their orange and black uniform jackets.

Schoutens had already prescribed individual protocols for the postal employees based on their schedules and chronotypes. Most work a set shift for an entire week before rotating, so the protocols attempted to change their clocks by a few hours for each weeklong shift, a "compromise" of sorts, and then help them quickly adjust back for the next. His primary clock-fooling tool was the glasses. But he shared other tips and tricks for sleep and meal timing.

CIRCADIAN CONSULTING, OR CHRONOCOACHING, IS A BURGEONING business. Police officers, firefighters, health care workers, pilots, astronauts, truck drivers, and athletes are just some of the people seeking help from Schoutens, Cheri Mah in San Francisco, and other experts.

Luke Gupta, a sleep and circadian scientist with the UK Sports Institute, worked with British athletes in preparation for the 2020 Summer Olympics in Tokyo. Just like Van Kerkhof, the athletes had to adapt to a substantial time change. Men and women competing in indoor sports, such as table tennis, badminton, and gymnastics, faced the greatest challenges. Once these athletes land at the airport, they "might not see the light of day for the entirety of the trip," he told me before the games. "You go to the airport, hotel, indoor training venue, back to the hotel, repeat." And not every team could afford to arrive weeks in advance to adapt naturally.

Gupta assisted athletes in speeding up their acclimation with "big doses of light" at the right time. He didn't enlist fancy glasses or light boxes. His focus was on natural light, outdoors or through windows. And the most effective tool to help them see that light, according to Gupta, was education. In the winter of 2018, he traveled with the gymnastics team to Tokyo. There, he fit the athletes with LYS light sensors and had them fill out "light diaries" as they lived and trained to help them discern between circadian-stimulating light and light that the system ignores—a difference that isn't always easy to see. When they traveled back for the main event, he hoped, the athletes would know better how to use the sun's power for optimal performance.

I met another chronocoach, Allison Brager, over coffee at the Society for Research on Biological Rhythms conference in Florida. I was immediately struck by the tattoo on her left arm: an artful depiction of the molecular structure of adenosine, the substance that builds up sleep pressure. As we spoke, our sips of caffeine made us more alert by temporarily appropriating the receptors normally reserved for adenosine and artificially muting the sleep signal. Caffeine is one of the tools that Brager enlists to support military special operations teams.

While giving soldiers high doses of a stimulant is not ideal, there is also no sense in trying to persuade the military to do away with shift work or extended operations, said Brager, a scientific officer with the US Army. "The army will never stop fighting at night." So she tries to help soldiers manage extreme circumstances as safely and effectively as feasible. Brager recalled being awake herself for seventy straight hours while on deployment in Kuwait. "I became very, very adamant about using that caffeine gum," she told me.

Brager also imparts her rhythm-hacking ideas to athletes competing in everything from weightlifting to professional hockey to video games. Brager works with the US Army Esports team, which, like any other military unit, often operates around-the-clock with rotating shifts. *Esports* is short for electronic sports, a form of competition using video games. The rapidly growing sport has become so popular that the team also now serves as national marketing for the army.

I spoke with Sergeant First Class Christopher Jones, who goes by the handle Goryn on Twitch and other gaming platforms. He founded and managed the team, which has grown to about eighteen thousand soldiers. Sergeant Jones told me that he had been a gamer most of his life, purchasing a PlayStation 2, GameCube, and Xbox as soon as he could afford to with military paychecks after enlisting. He played in his off time, generally late into the night, and even when deployed in combat zones. "Me and my buddies played *Halo* against each other, or whatever else we could get our hands on, in between missions," he said.

Now, playing equals training. Athletes on his team spend an average of sixty hours a week at it. Competitions take the team around the world, forcing them to confront the same challenges with jet lag and lining up peak performance times as athletes in other sports. Plus, their competitions tend to go longer: upward of eight hours for one game. Given the other commitments for soldiers in the army, it is not uncommon for his team's sleep to be sidelined.

Most members of Sergeant Jones's original team were night owls with peak performance times in the later evening. His own peak occurs around 7:00 p.m. That is rare for athletes. Most traditional sports teams tend to be a mix of chronotypes, often with more larks than owls at the elite levels. Brager encourages teams to turn that variability into an advantage. A team might determine each player's chronotype and corresponding peak performance times for that sport and then organize game rosters accordingly. Maybe that involves reserving a night owl baseball pitcher for night games or playing people with different circadian peaks at their uniquely optimal times over a four-hour football game, or an even longer cricket match. A team could even use this knowledge in forming their team. Authors of one study suggested that selecting more night owls for evening competitions, such as Champions League matches in soccer, or European football, could bump a team's peak performance later and up their odds of victory. Brager added her hope that the military would consider similar chronotyping in their operations and planning. Of course, a player's or soldier's baseline talent, strength, and stamina probably remain more significant factors to consider.

Individual or non-team sports can generally sidestep this complication. But there's still that challenge of lining up an athlete's peak time with the set competition time. The average peak performance for a swimmer, for example, is around 5:00 p.m., plus or minus a few hours for a true early bird or night owl. Race times tend to be significantly slower in the morning, by tenths of seconds. And, as we've noted, these same races can be decided by one hundredth of a second.

FLORENT MANAUDOU'S BEARD THREW ME OFF. HE HAD THE BROAD shoulders, athletic haircut, and confident gaze of a world-class swimmer but, I knew, even trimmed as it was, that much facial hair would add precious hundredths of seconds to his race times.

We met via Zoom in January 2023, on a week between competitions. Since his first Olympics in 2012, Manaudou had won four medals, including two gold, for his home country of France. One hundredth of a second was all that had stood between him and a third gold. The 2024 Paris Olympics would be his last attempt to add to his collection. "I'm an old sportsman," Manaudou told me.

At age thirty-two, he was seeking every advantage possible—however slight. That now included the chronocoaching of Schoutens, a pair of Propeaq glasses, and, I'm assuming, shaving his beard. Manaudou's peak swim time matches that of most swimmers: between 5:00 p.m. and 7:00 p.m., depending on the season. "When the sun goes down at five, it's not the same as when the sun goes down at nine," he told me. So far, he's used the glasses to counter jet lag and energize before training or races, especially during the dark winter months. But Manaudou most values the glasses in helping his body recover, something he has found more difficult with age.

He said other circadian tricks, like meal timing, have paired well with his glasses. In France, dinner can be late. Manaudou used to eat his around 9:00 p.m. But he told me he has veered away from that cultural norm and tries to finish by 7:00 p.m. After dinner, he dons his Propeaqs with orange-tinted lenses and goes about the rest of his evening: watching Netflix, using his phone or iPad, entertaining friends. They do give him funny looks for wearing the glasses. But Manaudou said he doesn't care. "I used to be up until 1:00 a.m.," he said. "Now I'm sleepy and going to bed by 10:30 p.m."

Another chronocoaching mantra: be consistent. Manaudou's sleep schedule once bounced by as many as five hours between training and off days. Handling that social jet lag was easier when he was younger. By adjusting his schedule to wake up and go to bed at the same time every day, Manaudou said, he's regained some of that youthful energy: "I have more rhythm."

The last stop on my tour through the Netherlands was an Olympic swimming complex in Eindhoven. Athletes from around Europe, including the French swim team, have come here to trim their times at the facility's InnoSportLab, which is staffed by PhD students and scientists who leverage tools to analyze factors such as vortices, propulsion, and drag. Underwater 3D cameras make much of this feasible. In a small room beside a teched-out training pool, I met Manaudou's coach, Jacco Verhaeren, the technical director of the French national swimming team.

Again, the smallest details can make all the difference for elite athletes. That detail might be the choice of swim cap, goggles, or Bono-esque blue-light glasses. "A hundredth of a second is the difference between nothing and everything," said Verhaeren, his voice raised so I could hear him over the bangs of "dry diving" boards in a training room next door. "No one wins because of glasses, but they can lose without them." In other words, a high-performance swimsuit or a pair of glasses will not turn a recreational swimmer into Katie Ledecky, but it could help someone like Manaudou turn back time a little.

The entire French swimming team now has Propeaqs, as do more than one thousand five hundred other Olympic athletes from sixteen countries competing in various events: skeleton, skiing, curling, judo, gymnastics, track and field, and more. In addition to individual athletes, Schoutens has consulted for team sports. He recalled working with the Dutch women's field hockey team ahead of its 2019 FIH Hockey Pro League title. The team had to travel over five continents in three months. Schoutens helped divide the team and staff into five chronotypes and ensured only people with the same chronotype shared hotel rooms.

Then there was Victor Campenaerts, a Belgian cyclist who wanted to break the world record for distance covered on a cycling track in one hour. His best chance of doing so was on a track at high altitude, where the air is less resistant. With Schouten's guidance, Campenaerts and his support team concocted a plan for him to achieve the feat in Mexico at about three in the afternoon, when his body would be near peak performance. But once they arrived, it became clear that it would be too hot at

3:00 p.m. "So we changed the hour to 11:00 a.m.," Campenaerts told me. This threw the cyclist for a loop. His body would not be optimally prepared to perform at 11:00 a.m. Campenaerts used a combination of the blue light in the early morning and the orange filters in the early evening to deceive his body into thinking it was hours later than it was. "I changed my whole rhythm," he said. "Meals, training hours, just my whole life was moved a few hours earlier in the day." Campenaerts broke the record that day, although other cyclists have since surpassed his mark.

Schoutens has also assisted two hundred people in a fifty-hour dance marathon, and a group of Dutch marines in a world-record-breaking speed march. Perhaps his most extreme case was DJ Giel Beelen. Schoutens helped the DJ break the nonstop radio world record. Tunable LED lighting simulated forty-two-hour days. Beelen persevered for 198 hours with two or three hours of sleep each artificial night and occasional short power naps.

CIRCADIAN HACKS CAN GET EVEN MORE WACKY. TAKE, FOR EXAMPLE, the bridge light beacons employed by Jamie Zeitzer at Stanford. During my visit to Palo Alto, he showed me how he manipulates people's inner clocks with artificial sequences of flashes during specific periods of their sleep. The technique, he explained, exploits the circadian system's pickiness about when it sees light. He pulled out one of these beacons, a clear cylinder about ten inches tall with a black base. Zeitzer's spiky and disheveled gray hair made him look as if he had just put his finger in an electrical socket. He told me he has only mildly shocked himself tinkering with the beacons. After being unplugged, they retain enough charge for one last flash. "You'd think after the first time I'd learn," he said.

The thing is bright, at least on the indoor light scale: about 1,000 lux. However, Zeitzer said he can still get results with beacons dialed down to a couple hundred lux. In a study of teenagers who struggled to fall asleep early enough to wake up for unnaturally early school start times, he programmed beacons to pump out three-millisecond flashes every twenty seconds during the last few hours before they woke up. The light penetrating their eyelids subconsciously advanced their clocks and made them sleepy earlier the next night. Combining the light with cognitive behavioral therapy earned the teens around forty-three minutes more sleep per night—significantly more than behavioral therapy alone, or the light therapy alone. Teens shown the light pulses without the behavioral component got tired but simply fought their biology, Zeitzer told me.

He has reset inner clocks by upward of four hours in one night. But individuals' responses vary widely. Why that is, he's not yet sure. More research is needed. Still, he anticipates exciting future applications. Flashes of light might help people preadapt to time zone changes ahead of travel. They could also help someone who stayed out a little too late dancing on Saturday night. Let's say you come home at 3:00 a.m. and sleep until 11:00 a.m. You have missed the morning light. Now that vicious cycle begins, pushing you to stay up late on Sunday night, too. But what if you need to get up at 6:00 a.m. on Monday for school or work? Zeitzer suggested that light could be automated to flash during your sleep, so even though you didn't wake up on Sunday until 11:00 a.m., your brain still caught the "sunrise." Being a light sleeper, I wondered how someone could sleep through these flashes. Zeitzer assured me that a similar proportion of participants say they saw the light whether or not they actually received the light pulses. It helps that only about 10 percent of light makes it through our closed eyelids.

The TUO light, the brainchild of Jay Neitz and James Kuchenbecker of the University of Washington, is yet another curious attempt to trick our clocks. By isolating ipRGCs from primate retinas and watching how their cells responded to lights turning on and off, and alternating between wavelengths, Neitz and his team zeroed in on a novel light recipe: blue and orange flashes of light, cycling nineteen times a second. Even though the light looks white to the eye and shines as bright as a typical light bulb, he told me, the TUO will do a superior job of stimulating the circadian system—comparable to that moving contrast of orange and blue light at sunrise.

The more passive the intervention, the more likely people are to use it. That is the motivation behind various other programmable and automated lighting systems. What if we could reset our clocks during a Netflix marathon? Samsung jumped on the bandwagon in 2023 with its launch of a TV boasting a "Circadian Rhythm Display" certification from VDE (Verband Deutscher Elektrotechniker), a leading electrical engineering certification institute in Germany. Samsung states that the TV's "Eye Comfort" mode automatically balances light exposure and color temperature to resemble natural outdoor light. Again, as we await greater regulation and proof of impacts, it's still the wild west for all these products.

The circadian market is booming, with no signs of slowing down. And leading science and innovation organizations have taken notice. The Defense Advanced Research Projects Agency, or DARPA, a research and development agency with the US Department of Defense, is currently developing what it calls the "ADvanced Acclimation and Protection Tool for Environmental Readiness (ADAPTER)." The aim is to create an implantable or ingestible bioelectronic device containing cellular factories that will release compounds on demand. Essentially it would work by remote control. Soldiers could use the device to rapidly recover from jet lag or adapt to shift work. It might also help them fend off bacteria that causes traveler's diarrhea. Christopher Bettinger, program manager in DARPA's Biological Technologies Office, anticipates that the platform could one day produce a "complex symphony of proteins and hormones" to improve circadian entrainment. A soldier could dial in whatever hormone and at whatever concentration they wanted, Bettinger told me.

Northwestern's Phyllis Zee is collaborating on the project. She hopes to assist in turning that living pharmacy into a system that can detect a person's circadian phase and appropriately time what gets delivered. "Although it sounds like science fiction," said Zee, "it probably is going to happen."

12

Circadian Medicine

"Good morning!" my surgeon greeted me, sounding almost annoyingly chipper. I sat reclined on a bed in the pre-op room—gowned, hair netted, attached to an intraveneous line. He inspected my left foot, which my care team had marked with a big "Yes" to hedge the risk of him accidentally slicing into the right (but wrong) foot. He gave me a confident smile as he recapped his plan to salvage my big toe. It was late in the morning, and I was months overdue for the surgery. I had postponed it once. But, in that moment, I wondered whether I would have been as confident going under the knife in the late afternoon, my scheduled time on that earlier date. Would my surgeon have been so alert and energetic? Would the nurses and anesthesiologists have been so on point? Needles don't always go into my arm that smoothly.

I was in the middle of finishing this book when I took the break for surgery. I'd been deep in the weeds of circadian science long enough that I couldn't help but ponder the implications as I prepared for the procedure: Did circadian disruption speed progression of my osteoarthritis? What is the best time to schedule the surgery? When during the day should I get my pre-op COVID-19 vaccine booster to ensure I wouldn't be forced to reschedule again? How might the coming anesthesia-induced nap affect my sleep homeostat and circadian systems? What hours should I be braced to fend off the worst of the pain with heavy post-op meds? What hours should I take anti-inflammatory drugs to reduce the swelling? The list went on.

Circadian science has something to say in response to each of these questions, even if not always a straightforward answer. For starters, yes, animal studies do suggest circadian disruption can accelerate arthritis. And procedures during late morning hours are, on average, more favorable than hours later in the day for a surgeon's sharpness and a patient's recovery from the trauma. (Although later in the day might be better for heart surgery.) Morning is apparently also preferred if you want to avoid a fatal anesthesiology error or an infection resulting from someone in your care team not washing their hands.

Just as there are best times of day to organize work and play, and to eat, the emerging science teaches us that there are optimal times for your organs and systems to process medicine, fight viruses, build immunity, and undergo surgery. Mounting data now describe how accounting for daily rhythms can influence a variety of outcomes—from the accuracy of blood pressure tests to the toxicity of radiation treatment to the effectiveness of flu vaccines. We now know that minding the clock could ease cluster and migraine headaches, the painful inflammation of rheumatoid arthritis, and recovery after a traumatic brain injury. It could also increase the success of lung and liver transplants and possibly even slow down the hands of the clock and add years to our lives.

There is still a lot to sort out. Yet given our rapidly expanding knowledge and biotechnology breakthroughs, a growing chorus of scientists and doctors say it's time to begin translating the science into clinics, hospitals, and home medicine cabinets. We can both optimize current medical treatments and identify promising new ones by exploiting these biological relationships. Meanwhile, laboratory research continues to underscore the striking nature of these connections.

ZOI DIAMANTOPOULOU ARRIVED EARLY ONE MORNING AT THE ANImal house in her lab in Zurich. "I was quite tired," she recalled. "I'm not an early bird at all." But Diamantopoulou, then a postdoc at the Swiss Federal Institute of Technology, got to work anyway. Her research focused on understanding cancer metastasis and, specifically, circulating tumor cells, cancer cells that break away from an original tumor and travel to other locations in the body. She had been counting these rogue cells in blood from mice and people with breast cancer, and had been continually puzzled by the vastly different numbers. Often, she and her colleagues would tally thousands of the cells per milliliter of blood in mice yet only a few cells in human patients.

But that morning, as she looked upon the dozing mice with envy, a realization struck: humans and mice live and sleep on very different schedules. Yet her team typically drew blood samples from people and the nocturnal animals during the daytime, while one was awake and the other sleeping. Could that disparity have something to do with the contrasting counts? Diamantopoulou set out to investigate, first by collecting blood from mice every four hours across the twenty-four-hour day. Of course, this meant she frequently had to spend day and night in the animal house. At least her disrupted sleep was rewarded with remarkable results.

Sure enough, as her team reported in 2022, breast cancer cells were much more likely to spread in a mouse's bloodstream while it was sleeping than while it was awake. And we are not talking about a slight variation. Levels of circulating tumor cells in the nocturnal animals peaked during the day at concentrations up to eighty-eight times higher than when the animals were active at night. Diamantopoulou and her team took blood samples from patients hospitalized with breast cancer, too. Nearly four times as many tumor cells circulated in patient samples collected at 4:00 a.m. compared with 10:00 a.m. The true around-the-clock variation is probably far greater. Most participating patients were in very early stages of the disease compared with the mice and, therefore, were presumably shedding fewer tumor cells. Plus, being limited to two daily samples in the hospital meant the study probably missed catching the actual highs and lows in roaming cell counts. The fact that they still saw such a difference, Diamantopoulou said, was impressive. It appeared that the tumor awakens while the patient sleeps.

Breast cancer is the most commonly diagnosed cancer worldwide. Any effective tactics to detect, halt, or slow its spread would be welcome. The findings hinted that exploiting the clock by, say, taking biopsies and dosing certain breast cancer treatments during the night, could potentially raise a patient's chances of survival. Diamantopoulou and her colleagues emphasized that the findings by no means suggest that cancer patients should avoid sleeping. Sleep itself is critical for recovery. Similar could be said for people with asthma.

IN 1869, HYDE SALTER, AN ENGLISH DOCTOR, WROTE HIS OBSERVATIONS of a pattern people had noticed for centuries: asthma tends to attack at night. He translated that phenomenon into advice for his fellow doctors—specifically on when to dose the era's go-to asthma medication, a nightshade called belladonna: "Giving it at night, you bring the full force of the drug to bear upon the disease at the time at which it is most liable to come on." Not only would that timing be more effective, he suggested, but it would also minimize the unwelcome side effects of the poisonous plant: "By giving it only once in the twenty-four hours, you are able to give a larger dose than you would be able to do if oftener repeated. By confining it to bedtime, the patient's days are, in spite of a large dose, passed in comfort; for, as the morning advances, the dullness of head, confusion of sight, and drought of mouth pass away."

More than a century and a half later, research confirmed that asthma symptoms follow a circadian rhythm with severity peaking at night for most people, independent of sleep and other daily cycles. In 2021, scientists echoed Dr. Salter's words, in more modern medical speak: "Short-acting drugs could be targeted to counter the circadian time of worst pulmonary function and reduce side effects rather than simply giving patients the maximum tolerated dose of medications across the entire day."

In addition to providing greater treatment options than a toxic nightshade plant, medical science today can also begin to explain the molecular underpinnings for time-of-day effects rather than rely on observations or trial and error. We now know that clocks tick in nearly every cell and organ system and that this orchestra of timekeepers constantly turns on and off a bevy of genes. We know that about half of our genes follow a day-night rhythm, with many genes—particularly those in the brain and testes—fluctuating with the seasons, too. A lot of these rhythmic genes encode the molecular targets for drugs, the proteins that transport them, and the enzymes that break drugs down. Research suggests that the majority of today's best-selling drugs, including treatments for asthma, arthritis, high cholesterol, and post-op inflammation, probably work better when taken at specific times of day. Yet a paucity of these drugs in the US carry FDA-approved recommendations for when they should be taken.

Pertinent to my surgery, I learned that taking nonsteroidal anti-inflammatory drugs, or NSAIDs—in my case, ibuprofen—during the day but not at night may be most effective for post-op healing. Inflammation follows a circadian rhythm. While daytime inflammation can be destructive for bone healing, science suggests that nighttime inflammation can be beneficial. Plus, NSAIDs may suppress melatonin. I needed sleep to heal, too. So the rule of thumb I followed for my big toe: anti-inflammatories during the day and pain-killing analgesics during the night. Nighttime, I also learned, is when pain tends to peak.

Similarly, the time when a person with diabetes takes metformin could affect how much the drug reduces blood glucose levels. Antiepileptics and other treatments that we may or may not want to get inside the brain could be timed to when the blood-brain barrier is more or less permeable, too. Indeed, I spoke with several experts who believe that many more FDA-approved drugs—perhaps even birth control pills—surely have time-of-day effects. And that optimal timing could be identified without too much trouble. Because these drugs have already been shown to be safe in humans, investigators could skip early phase trials. "Can we make drugs work better? It looks like the answer is yes," John Hogenesch, the chronobiologist at Cincinnati Children's, told me. "In some cases, two times better than they did before." By hitting a drug target at the right time of day, there's an opportunity to drastically reduce side effects, too.

The COVID-19 pandemic boosted momentum for circadian medicine. Researchers discovered that timing mattered: jabs with an mRNA vaccine appeared most effective between late morning and early afternoon—the window I targeted prior to my surgery—and PCR tests were least liable to miss detecting an infection around 2:00 p.m. There were caveats, like a lack of data points for late-night jabs and a lack of accounting for individuals' internal timing. And, of course, simply getting vaccinated and tested is always of greater importance than when we do it. Adequate sleep before vaccination improves effectiveness, too. Scientists further found that circadian disruption put people at greater risk of getting COVID-19, and that people with long COVID showed symptoms of circadian disruption.

None of this should have surprised us. After all, the immune system is one of the most circadian of the body. Virtually all its components multiply, migrate, and wage war according to the clock. We likely evolved these patterns so that our defenses—from the impermeability of our skin to activation of our immune cells—could peak when the body was most susceptible to be attacked by bacteria, viruses, parasites, or even cancer cells. Now, scientists are using this intel to suss out what's happening both when our immune system is fighting foes and when it is screwing up and attacking our healthy cells, such as in autoimmune disorders or in the cytokine storm, an overreaction of the immune system that has led to severe forms of COVID-19. And scientists are listening to the ticking of not only the clocks in our cells but also the clocks in cancer cells and even in the living cells of threats outside our body. It can be quite the cacophony.

EVERY YEAR, MALARIA SICKENS HUNDREDS OF MILLIONS OF PEOPLE. The global scourge claims the life of a child under age five nearly every minute. To make matters worse, malaria-carrying mosquitoes are swarming into new areas that are warming with climate change, expanding the reach of their piercing proboscises and the parasite that causes malaria.

An infection with that parasite, *Plasmodium*, will trigger fevers on an incredibly regular schedule, always in a multiple of twenty-four hours. Hippocrates observed these oddly recurrent fevers nearly 2,500 years ago: "When the paroxysms fall on even days, the crises will be on even days; and when the paroxysms fall on odd days, the crises will be on odd days." Proposed explanations during his time included the presence of "demons." Modern science has since deduced that malaria's rhythmic fevers are instead a result of parasites' synchronized replication within the host: multiplying within red blood cells and then bursting open those infected cells in unison to release newborn parasites into the bloodstream and begin the cycle anew. Demons still seem an apt depiction.

It behooves the parasite to deliver its progeny when the host's blood is well stocked with nutrients that new parasites need to thrive—like right after the host eats a meal. Meanwhile, the host will put up its defenses to avoid sharing its resources. "It's not easy living in something that's trying to kill you all the time," said Sarah Reece, an evolutionary parasitologist at the University of Edinburgh. She and others have discovered that parasites do particularly clever things to dodge dangers and acquire food, like telling time.

Scientists had assumed that malaria's predictably periodic patterns of infection were enforced by the circadian rhythms of the human host, and that the parasite was simply along for the ride. We now know the microscopic little demons have rhythm, too. They can entrain that rhythm to their host, much like how we sync our clocks to the sun's cycles. In a 2023 study, researchers collected blood from ten people with malaria and found that the activity of hundreds of *Plasmodium* genes ramped up and down in coordination with the daily rhythmicity of the patient's circadian clocks. Once their rhythms are tethered, a parasite might take advantage by scheduling its rupture of red blood cells to when it can steal nutrients from its host. But hopefully now we can, in turn, exploit this understanding and one-up them in the arms race.

The primary treatment for malaria is artemisinin in combination with partner drugs. Each dose of artemisinin lasts only a few hours in the body, giving it a small window of opportunity to kill parasites. And the parasites vary in their susceptibility to the drug—and to every other antimalarial thrown at them—based on where they are in their replication rhythm, Reece explained. While the science is not yet prepared to make specific recommendations, thoughtfully timing treatments could theoretically make the drug more effective and could mean getting away with doses less toxic to the human. A greater understanding of this nesting doll of clocks within clocks might also inform the development of novel drugs that target various aspects of the dynamics between parasite, host, and vector. The latest research even hints at the possibility of new antimalarial drugs that could throw a wrench in *Plasmodium*'s clockwork, essentially jet-lagging the parasite or decoupling its rhythms from its host. Such strategies may be effective not only for malaria but also for sleeping sickness and other rhythmic parasitic diseases that afflict hundreds of millions of people around the world.

Of course, *Plasmodium* appears to organize its rhythms to fit the rhythms of not only its human host but also its mosquito host, which serves as its vector to us. The parasite can replicate all it wants, but its real goal is to find more hosts to infect. And it relies on the much-maligned mosquito. Studies show that mosquitoes are more prone to get infected when exposed to *Plasmodium* during the day as opposed to at night. But here's a twist. Historically, malaria-transmitting mosquitoes feasted at night and slept during the day. In some regions, however, mosquitoes are modifying their rhythms to bite at the fringes of daytime—either earlier in the evening or toward dawn. This might be a move to evade insecticide-treated bed nets. What it means for transmission is not yet clear. If the changes in foraging times make mosquitoes more susceptible to contracting malaria, that's obviously concerning. But Reece offered an optimistic caveat: because the mosquitoes are messing up their internal rhythms by changing mealtimes, just as we modern humans tend to do, they might not be the reliable vectors they once were for parasites. *Plasmodium* needs about twelve days in a mosquito before it's ready to go back into a human. That is already a long time for a mosquito to live, and potentially unachievable with mangled rhythms. Fewer mosquitoes may live long enough to transmit malaria. "That is the hope," said Reece.

What might be even better than shortened lives for deadly mosquitoes? How about extending our own longevity? Circadian scientists are working on that, too.

THE TICKING OF OUR CLOCKS BEGINS AND ENDS QUIETLY. WE LIKELY start with no internal rhythm during at least our first weeks in the womb. Then the circadian system begins to set up its clock shop. If all goes well, our clocks will collectively tick loud and strong well into adulthood before they slowly weaken during the last decades of life. Interventions to bolster the building and repair of clocks at both ends of the life span could have profound benefits for our health and well-being.

In addition to novel lighting in the NICU and thoughtful timing of breast milk and intravenous feeding, Hogenesch and his colleagues are implementing more clock-supporting strategies at Cincinnati Children's. In 2021, they opened a circadian medicine clinic, the first such center in the country dedicated to childhood circadian rhythm sleep disorders. The team is discovering strong links between circadian clock issues and neurodevelopmental conditions including autism. Clock repairs early in life may reduce these problems, Hogenesch told me. He shared the story of Jack, a teenager with Smith-Kingsmore syndrome. The rare neurodevelopmental disorder is caused by a mutation in a single gene. Before his team intervened, Jack would engage in repetitive movements known as stimming about 350,000 times a week. He also had severe sleep problems. A drug Jack was taking to turn off the implicated gene had untethered his body from the twenty-four-hour day-night cycle, casting his clocks adrift similar to a blind person who lacks melanopsin. After the clinic team adjusted his dose and added a new sleep aid, Jack's sleep dramatically improved. He also dropped his stimming to 50,000 times a week. Comparable circadian issues accompany autism, Down syndrome, genetic epilepsy, Rett syndrome, and more, suggesting additional opportunities for improvements, according to Hogenesch. As he pointed out, a newborn has many more neurons than a five-year-old. Treating young kids during this early period of "pruning," he said, may avert intellectual disability and developmental delay.

Ticking troubles return later in life for almost all of us. As we age, our rhythms dysregulate and dampen, which scientists now think could be a common cause of age-related diseases, including hearing and vision loss, pulmonary diseases, depression, dementia, metabolic syndrome, heart disease, and osteoarthritis. Clock changes are especially pronounced in people who develop Alzheimer's disease, which is predicted to affect nearly fourteen million Americans by 2060—barring further treatment breakthroughs. Circadian and sleep disruptions tend to appear before the onset of dementia-related memory loss. Then they worsen during the course of the disease. Patients nap at all hours. Day and night lose their distinction. It's a major driver for needing residential care.

The proverbial chicken-or-egg question naturally arises: Does neurodegeneration cause this blunting of rhythms, or does the blunting cause neurodegeneration? If it's the latter, could we rescue weakened rhythms and prolong the quality and quantity of our lives? "There are a lot of question marks," Erik Musiek, a neurologist at Washington University in St. Louis, told me. But he and other scientists are collecting clues. We know that both long-term shift work and chronic sleep loss in midlife may slightly raise the risk of Alzheimer's later in life. We know that the immune cells that clean up beta-amyloid plaques follow a circadian rhythm. We know that the activity of neurons in the suprachiasmatic nucleus decreases as we age. And we know that removing cataracts and, therefore, eliminating a roadblock for blue wavelength light to reach the ipRGCs at the back of the retina may decrease dementia risk.

We should know even more soon, perhaps by the time you're reading this book. Musiek and his team have been following a large group of older adults for more than a decade, tracking their daily routines and watching their brains for the development of beta-amyloid plaques and dementia symptoms. "We'll be able to see which comes first—rhythm problems or plaques," he told me. Various research teams are also tinkering with molecular clocks in mice via genetic manipulation or abnormal light-dark cycles to see how that affects the hallmarks of the disease.

Circadian rhythm disruption and neurodegeneration may interact to form a deleterious cycle. It could be that broken clocks are both a consequence of and a modifiable risk factor for Alzheimer's disease, and that intervening early could stall the spiral. Even if a person has lost natural clock function in their cells, manipulating their environment with light, food, exercise, or other means might "fake the clock," as one scientist put it, and induce enough rhythmicity to clear out plaques. Until we have more answers, Musiek tells his Alzheimer's patients to get outside in the morning, keep a consistent sleep and meal schedule, and be diligent about not eating or turning on lights late at night. He's hopeful that this basic circadian hygiene will help slow progression of the disease or at least lessen the severity of symptoms.

Shifting, synchronizing, and strengthening our rhythms might even soon be as simple as popping a pill. Scientists have identified compounds that may one day prevent or treat Alzheimer's and, more generally, extend our lives. Or they might just make our lives easier.

IMAGINE FLYING FROM NEW YORK TO LONDON, OR EVEN BEIJING, and never experiencing the malaise of jet lag. Cordycepin, a compound derived from a fungus, has induced clock changes of up to twelve hours in humans and mice. Also intriguing, a phytochemical compound that cumulates in the peel of citrus fruits, called nobiletin, has prompted sustained increases in clock amplitude in mice. These animals showed improvements in several measures of healthy aging, including energy, metabolism, sleep, and skeletal muscle function. Much more research is needed before we know if it is effective, or even safe, in humans. We do, however, know that a diet rich in fruits and vegetables helps fend off a range of age-related illnesses. Could this circadian effect be among the reasons why?

Many more clock-shifting and clock-enhancing candidate compounds are under investigation. Some are drugs already widely used in medicine for chronic conditions, such as rapamycin. Others are common supplements, such as folic acid. Others have yet to earn a name. Carrie Partch and her team in Santa Cruz are investigating a small molecule that can directly bind to clock proteins. By enhancing our circadian machinery's ability to make the right moves at the right time, a compound could promote healthier aging. It could aid in acute scenarios, too. Resetting clocks by up to twelve hours could enable the military to land personnel anywhere in the world ready to aid in disaster recovery. Or imagine a pedestrian got hit by a car at 5:00 p.m. While there is never a good time to be hit by a car, 5:00 p.m. is the wrong time. Research suggests that daytime wounds heal faster than nighttime wounds. So, what if when they first arrive at the hospital, doctors gave a drug to dial back their clock to 8:00 a.m.? Or what if I had kept that first late-afternoon surgery time and simply popped a pill to jet lag my internal time, artificially rescheduling the surgery to an earlier time?

Drugging our clocks may also make some drugs work better. Michael McCarthy, a psychiatrist at the University of California, San Diego, has studied the response of bipolar patients to the mood-stabilizing drug lithium. Despite only about one in three patients responding well and long term to the medication, it's remained a primary treatment option for several decades. When it works, it works. But what determines whether it will work for a particular patient? That question has confounded doctors. Now, scientists are finally uncovering clues. McCarthy has found that night owls and patients with weak circadian rhythms tend to respond poorly to lithium. And he thinks he knows why: Lithium lengthens and strengthens circadian rhythms for those who respond well. So a patient with a naturally short period might be better able to tolerate its period-lengthening effect. His data collected from neurons grown from bipolar patients so far support that logic.

More generally, McCarthy has found that bipolar patients have weaker circadian rhythms than the general population. He told me that rhythm is "almost entirely absent" in patients who fail on lithium. Whether or not lithium's impact on the circadian system explains the treatment's therapeutic effects or if it is just part of a more complex process remains to be seen. If alterations to the molecular clock alone are therapeutic, then strategies that support circadian rhythms should benefit people with bipolar disorder, McCarthy said. In fact, studies already show light therapy can improve symptoms. There could be synergistic effects, too: one study found that light therapy and sleep deprivation—both known to be effective in treating depression—improved the efficacy of lithium.

McCarthy and his team have also looked closely at genes implicated in bipolar disorder and saw signs that they regulate rhythms. And other research finds certain antidepressants can make the circadian system more sensitive to light. They might literally help a patient look on the bright side, but only if a patient taking the drug is exposed to healthy light patterns. "So many different things can be going wrong in psychiatric disorders," said McCarthy. "Maybe the circadian clock is the glue that holds all of these things together."

DRUGGING CLOCKS AND CLOCKING DRUGS COULD BECOME NEW weapons in the fight against cancer. Scientists hope to add both tactics to the arsenal against glioblastoma, among the deadliest cancers. "Glioblastoma is such a terrible disease," said Erik Herzog, a chronobiologist at Washington University in St. Louis. "These patients are living with a fifteen-to-twenty-month prognosis and a really rough ride along the way." He didn't need to convince me of this cancer's evil. It took my mom's life.

Clocks in tumor cells had been thought to rarely tick in tune with clocks in healthy cells. In fact, clocks are strongly altered or even broken in many cancers. But in the case of glioblastoma, the cancer's robust daily rhythms may participate in accelerating its growth and spread. This discovery has led to early investigations of clock-drugging candidates that may one day put the brakes on glioblastoma and slow the tumor's lethal rampage. Meanwhile, other research is focused on taking advantage of the windows of opportunity afforded by glioblastoma clocks.

In one experiment, Herzog and his colleagues engineered cells from patient tumors to express luciferase—the protein that makes fireflies glow—every time core clock genes switched on in tumor cells. Then they watched. "It was so dynamic," one of Herzog's colleagues told me. "Lights go on, lights go off. Lights go on, lights go off every day." Then the team started treating the tumor cells with drugs at different times in the cells' daily cycle and found that they were most sensitive to an oral drug, temozolomide, near the daily peak in activity of a core clock gene. Their discovery suggested that if doctors could direct patients to take this pill—part of the standard glioblastoma treatment—at this time of peak expression, the drug might be more effective.

They tested their theory in a retrospective study and found that people who had been treated with temozolomide in the morning lived an average of three and a half more months than those treated in the evening. The survival boost was significant, Herzog told me. The FDA granted approval for the drug based on studies that determined it could extend survival by two and a half months. A prospective trial is now underway, the first clinical trial in the US to apply circadian medicine in cancer. So far, compliance is high with no notable differences in side effects. More time will tell if timing extends survival.

DESPITE THE POTENTIALLY PROFOUND IMPLICATIONS OF CIRCADIAN medicine, it remains at the fringes of clinical practice. It is scarcely recognized or applied by doctors, let alone patients. The reasons for that, experts told me, are multifold. They include messy and incomplete data. They include concerns over cost and convenience. And they include a dose of ignorance.

Medical schools pay scant attention to rhythms. Take, for example, the highly ranked Johns Hopkins University School of Medicine. Chi Van Dang, the scientific director of Ludwig Cancer Research at Johns Hopkins, told me that neither circadian science nor sleep biology gets more than a cursory mention in preclinical courses. As a result, he said, few practicing clinicians know that a drug's effects can depend on the time of day. "If you take aspirin in the evening, it is a hell of a lot better than taking it in the morning," said Dang, referring to the drug's use in low doses to prevent blood clots and reduce blood pressure. In one study, patients who received aspirin during breakfast had increased risks of hemorrhage, cardiovascular disease, and death. How many doctors recommend nighttime dosing to patients? Dang's guess: about one in twenty.

Medical practice and medical school curricula rest on a foundation of scientific research—from studies of laboratory animals through multiple phases of clinical trials. Only a tiny fraction of such studies and trials have considered a treatment's timing. Even in 2023, as I write this, less than one of every thousand cancer clinical trials currently incorporate circadian-timed treatments. These omissions have led to potentially great losses for medicine. Time and again, drugs that have shown stellar promise in mice and rats fall flat in clinical trials with humans. It's a theme across all drug development: more than 80 percent of novel drug candidates fail when they're initially tested in healthy volunteers and patients. More fail in subsequent large-scale clinical trials.

Eng Lo, a neuroscientist at Harvard Medical School, and members of his lab grew curious about whether the opposing circadian cycles of rodents and humans could be a significant contributor to those failures. Most scientists conduct studies during the daytime, often in the morning, when rodents sleep. Lo decided to dig up a sample of stroke drugs that had appeared promising in animal studies but failed in clinical trials of stroke patients. He tested animals at the rodent equivalent of the human morning—when stroke patients had been treated. This time, the drugs didn't work well in rodents either. That raised a provocative possibility: Could the drugs be effective in humans at the equivalent of the rodent morning? By this time, the development programs for these stroke drugs had been abandoned.

Many factors differ between a rodent model and a human beyond just sleep timing. The animals typically don't come with comorbidities or older age, for example. Still, Lo suggested that timing is vital. "In my heart of hearts, I think this is an important reason that gets no airplay, and it's a fairly low bar to pay attention to," he said. "Whatever smart research we do that is based on molecular targets, we have got to think about time of day."

Adding the dimension of time to a clinical trial, or even to basic research, is not necessarily simple. A typical clinical trial for a drug today might give a bunch of people a placebo and a second bunch of people the drug of interest, follow them all for a year or so, and then see what happens. Factoring in time of day would mean giving the placebo to multiple groups of people, each at different times of day, and giving the drug itself to yet more groups of people, each again at different times of day. This adds complexity and cost. The price tag of the trial can quickly ramp up, on top of the logistical and statistical headaches. "It's just unattractive for all kinds of reasons," said David Ray, a circadian scientist at the University of Oxford.

Ray noted another challenge. Imagine a pharmaceutical company identifies a narrow "best" window for its new drug, say, in the morning, and formulates it with a higher dose because they now know it will pose less risk of toxicity in the morning. "But what if someone takes it at night by mistake?" said Ray, noting that few of us actually read or listen to our doctors' instructions. Regulators could ensure that marketing for a medication optimized for a specific time of day carry extra warnings about the risks of deviating from the schedule. But then that's not a good selling point for a liability-wary drugmaker.

The trend among pharmaceutical companies is actually in the opposite direction: to concoct once-a-day and other long-acting drug formulations. Scientists worry about the consequences here, too. Sustaining levels of a rheumatoid arthritis drug that targets the inflammatory molecule TNF-α, for example, could leave the immune system unnecessarily impaired throughout the day. You only really need to block the inflammatory molecule for a four-to-five-hour window, noted Ray. Plus, the way companies design long-acting drugs can make them hard for the liver to metabolize and result in yet more toxicity.

It might just take one or two big wins by pharmaceutical companies—showing a dramatic reduction in side effects or improved effectiveness by timed delivery of a drug—and then every company testing drugs will want winners, too.

Toward that end, drugmakers could also get extra clever. Maybe they explore a time-of-day effect in preclinical studies and then use that information in the design of human trials. "We have vast amounts of information about which genes oscillate in which circumstances in which cell types," said Ray. Such intel on a drug target's daily patterns could pinpoint the best windows to dose the drug in a clinical trial, limiting the number of needed treatment arms. A more effective drug also shouldn't require as many trial participants to show that it is statistically superior to a placebo.

If drug companies needed further incentive, added Ray, they could devise ways to formulate their drugs for timed delivery and, therefore, prolong the product's patent protection. They could even go further and automatically couple timed drug administration with an individual's clock. Ray envisions a timed-release capsule activated by an internal time cue, such as core body temperature or levels of oscillating hormones. Until that technology is ready for prime time, a person could still receive an alarm reminder to take drugs at their optimal time. "I think it's all eminently feasible," he said.

Circadian scientists offer another heavy-handed option: require that researchers factor time into their government-funded studies. The National Institutes of Health now does this for "sex as a biological variable."

Even if a time-of-day effect has been identified and acknowledged, applying circadian medicine is still seldom trivial. Hospitals tend to dole out drugs during hours convenient for staff. Team rounds and shift changes are the most typical treatment times. In one analysis of drug distribution in a hospital, researchers highlighted a telling example: while an antihypertensive medication, hydralazine, is most effective at night, care teams administered fewer doses at night than at other times of day. Patients also sense pain more acutely in the evening yet are usually given pain medications during morning or afternoon rounds in the hospital.

Meanwhile, at home, we may be unaware that the time we swallow a medication could matter. Even if we knew when a drug should be optimally taken, that doesn't guarantee that we would pop the pill at that time. Only about half of people with a chronic illness follow their treatment recommendations. And what if the time suggested is two hours after a person usually falls asleep, or before they wake up? Good luck with that. It's a similar story when we need to schedule a procedure or surgery: many of us probably take the most convenient or earliest available time slot. That's what I'd originally done. A circadian-timed surgery or procedure is even thornier than medications. A surgeon can't cut into everyone at the same time. And seats in a chemotherapy unit can book up in much the same way as those for a movie.

But what if you didn't need a seat? Just as streaming movies from our homes has shortened lines at the theaters, transportable and automated infusion pumps are cutting our tethers from the hospital. What was once science fiction is, again, quickly turning into reality. Advances in time-released drugs and treatment pumps, such as for insulin-dependent patients with diabetes, have already revolutionized medicine and could be further harnessed.

CAROLE GODAIN REMEMBERS THE LITTLE DETAILS FROM THE CLINIcal trial she took part in more than a decade ago. There was the blue button she pushed to get her chemotherapy drugs, and the green light that came on to confirm that the medication was dripping into her veins. Then, of course, there was the hour—10:00 p.m. without fail, for every treatment.

By all accounts, Godain's own time was running short. The first treatment for her colorectal cancer had failed, and her last body scan had revealed twenty-seven tumors growing inside her liver. The psychologist from Tours, France, jumped at the opportunity to take part in a trial to test whether delivering drugs at a specific time of day might make them more effective or reduce their toxic side effects. Ideally, it would accomplish both. And, for her, it did just that.

Godain is cancer-free today. Francis Lévi, the oncologist who treated Godain, said that although such a fantastic result is anomalous, emerging evidence should encourage more interest in leveraging circadian medicine. One of the most cited studies of its application in cancer was published back in 1997. Lévi, also a chronobiologist at Warwick Medical School in Coventry, and his team had randomly assigned 186 people to either timed or standard treatment for metastatic colorectal cancer. Slightly more than half of the people who, like Godain, had their chemotherapy infusion synchronized with their circadian rhythms responded to the treatment, compared with 29 percent of individuals on a standard schedule. Her group also experienced fewer side effects.

Yet many trials trying to reap the benefits of timed drug delivery have had more equivocal findings. Most patients haven't been as fortunate as Godain. In a later, larger trial led by Lévi, more than five hundred people with metastatic colorectal cancer received specially timed or conventional chemotherapy. Survival times proved similar in both groups. But when results were broken down by sex—there's good reason the NIH is pushing this—the risk of an earlier death dropped by 25 percent for men. In women, it increased by 38 percent.

Such complexities lead Aziz Sancar, the biochemist and Nobel laureate, to be circumspect about applying circadian medicine to cancer treatments. "We all want it to work," he told me. "But we're not there yet." Sancar has enough hope that he is continuing his research on the optimal timing of cisplatin, a common anticancer drug. Years ago, he discovered the repair of DNA damage caused by cisplatin was clock controlled. "I will keep trying," he said.

There are a lot of overlapping factors at play in addition to sex differences. Optimal timing also appears to depend on the specific type and location of the cancer, as well as the phases in the circadian rhythms of both the patient's cells and the cancer's cells. Researchers found an uptick in circulating tumor cells during the day for multiple myeloma, for example, as opposed to the nightly rise that Diamantopoulou discovered for breast cancer. The bottom line is that there are likely no one-time-fits-all solutions in circadian medicine.

Before Godain started her home regimen, she strapped on a watch-like device that logged her daily rhythms. It detected that she had very regular sleep-wake cycles, which Lévi thinks probably contributed to her successful treatment. He and his colleagues have found that the median survival time of patients with weak rhythms is about half as long as those with strong rhythms, independent of other factors. Lévi is now testing a telemonitoring system in patients with pancreatic cancer. The device, embedded in a T-shirt or bra, will measure activity and temperature and transmit that data via Bluetooth to health professionals.

As Lévi emphasized, circadian treatment must be dynamic with a patient's unique and changing rhythms. Cancer treatment itself can alter the patient's clock. And the cancer's clock could change, too. Considering all these clocks adds further complexity but ultimately remains doable, according to Lévi: "We have tools we didn't have before."

TIME COULD BE THE NEXT DIMENSION OF PERSONALIZED MEDICINE. Yet prerequisite is the ability to read the time on our internal clocks. Determining the state of the circadian system, with its trillions of microscopic timekeepers, is difficult. A chronotype questionnaire is probably insufficient to precisely time a treatment. But scientists have come up with ways to make more educated estimates of circadian phases based on the timing of biological processes under the system's control.

Dim light melatonin onset, or DLMO, is the current standard. But I can tell you from experience that the process is time intensive and far from fun, especially for anyone who struggles—like me—to produce loads of spit on demand. It's also costly. Experts are scrambling to find simpler, faster, and less invasive methods than DLMO. They are also looking for ways to read the peripheral clocks throughout the body. Melatonin provides only an estimate of the master clock's time. That may not tell you what time the liver or muscle clocks think it is. And it probably won't tell you what time a tumor cell thinks it is either.

From a blood or hair sample, or even a micropunch of skin or a microneedle placed under the skin, scientists aim to extract the expressions of time-telling genes and measures of circulating metabolites. Each might rise and fall at different times during the day. Look at enough of them, and you can get an accurate prediction of internal time. The idea is similar to the Linnaeus flower clock, with its different flowers opening at different hours of the day. Line up a lot of flowers, and you can ballpark the time.

The least painful time teller I tried involved a research-grade smartwatch that tracked my activity, sleep, and light exposures. Many popular wearables, as well as more sophisticated technologies in the works by DARPA and others, could help determine the real-time status of our clocks as well as provide a treasure trove of data for users, their doctors, and researchers. Scientists, including Benjamin Smarr in San Diego, are already showing the utility of collected measures. Patterns over time in temperature, sleep, and heart rate, for example, may predict things like the onset of COVID-19 and pregnancy.

At this point, however, the tools haven't been widely integrated into large clinical studies. We may be missing not only an opportunity to inform timely medical treatments or zero in on the most effective therapy but also a chance to detect desynchrony and point to a prescription for fixing the clock itself.

Some of the most effective tools remain the simplest and cheapest. Tracking our rhythms and following the basics of circadian hygiene—optimally timing our light exposure and meals, in particular—could help stave off disruption and its associated health problems. Moreover, wellness and lifestyle changes don't need to be approved by the FDA. As we pursue futuristic treatment tools, we should each still aim to postpone their necessity as long as we can. I'd certainly prefer to avoid further joint surgeries. Prevention is the best medicine. The same could be said for efforts to sustain life on Earth, even as humanity develops the futuristic tools to reach faraway planets.

13

Extended Hours

In the mid-1990s, around the time a small Mars rover named *Sojourner* began rolling around on the red planet, Andrew Mishkin approached a watchmaker with an unusual request. He wanted to slow his black twenty-four-hour-face Zodiac Hermetic watch just enough so that it would complete its round in twenty-four hours, thirty-nine minutes, and thirty-five seconds—the length of a Mars day, better known as a sol. "It's slowing down the watch way more than any watchmaker would normally be willing to do," he told me. Fortunately, the watchmaker obliged. For the next couple of decades, Mishkin would wear two watches—one on Earth time and one on Martian time. He eventually retired the Zodiac in favor of a Martian-time-telling smartphone app.

Name a Mars rover mission, and Mishkin was probably part of it, if not commanding it as a systems engineer at the Jet Propulsion Laboratory in California. During many months of those missions, he and his team needed to live and work according to the time it takes Mars to rotate on its axis, as opposed to the twenty-four hours it takes Earth to do the same.

An extra forty minutes on the daily clock may sound like a good thing at first. There is rarely enough time in the day to accomplish everything on our to-do lists. But imagine the jet lag endured by crossing two time zones every three days. As those bonus hours accrued, rover crew members found themselves working the night shift and then the day shift again, their clocks left scrambling. A crew member once proposed they put mission control on a cruise ship following the sun so that the light-dark cycle would more closely match Martian time. "That never went anywhere," Mishkin told me. "But it was an interesting concept."

NASA has implemented a few more practical strategies, however, such as blackout curtains and special lights to help the Earth-based crews acclimate to the long Martian day, just as they have for the ninety-minute day—including sixteen sunrises and sunsets—on the International Space Station. The agency is also prepping to help astronauts and possible future colonies of civilians survive life on the red planet itself. Volunteers already occupy a simulated Mars habitat, part of a series of yearlong experiments leading up to the anticipated real thing. Meanwhile, Earth-based crews continue to explore Mars via remote-control rovers.

The extraterrestrial day lengths have been least welcome for the early birds on Mishkin's teams. The Mars sol is simply a better match for people whose clocks run a little longer. In fact, those extended hours might suit those at that end of the chronotype curve better than an Earth day. As NASA astronaut Kjell Lindgren suggested to me, we may "need to hire a bunch of night owls" for future missions to Mars.

Of course, life on Mars would force our bodies to contend with more than just a day length difference. There would be the planet's harsh environment, with chilling temperatures, barren red sands, and whipping winds. There would be the isolation, confinement, and monotony. There would be the intense levels of radiation and carbon dioxide, and the weak gravity, magnetic field, and atmospheric pressure. These changes could affect our physical and mental health, and further fool our internal clocks. Even more critical to our circadian rhythms, we Earthlings would be unaccustomed to the alien quantity and quality of daylight—which sounds uncomfortably reminiscent of my bunker environment. The sky on the red planet is dimmer and, well, more red. Very little blue reaches its surface. So natural Martian lighting would not sufficiently stimulate the circadian system. We would, at the very least, need to supplement with more of those tunable LEDs.

That's just the beginning of the lengths we'd need to go to survive on Mars. Imagine transporting food, water, and fuel to support people and ecosystems millions of miles from Earth. We struggle enough with the economic and environmental costs of shipping fruits and vegetables a few hundred miles. During his first NASA mission to the International Space Station in 2015, Lindgren was part of the first US crew to grow and eat a crop in space: "Outredgeous" red romaine lettuce. Then, during his 2022 mission, he worked on an experiment investigating the use of soil-free systems. "Can we grow plants in space? The answer is yes," he said. "How can we scale that up to actually provide a consistent, reasonable food source? That's the next step."

Circadian science could assist in this endeavor, too. Like us, certain crops have longer circadian rhythms that might resonate closer to Martian time; others, such as domesticated potato and carrot, might struggle. Rather than be picky about our veggies, we could use breeding and gene editing tools to fine-tune plant clocks. We could also manipulate day length itself with indoor agriculture fitted with LED systems. Lindgren said he envisions a future greenhouse module in which food and medicinal plants are grown as part of a sustainable system that, along with other feats, scrubs carbon dioxide from the atmosphere. Mark Watney and his Martian potatoes loosely portrayed aspects of this science fiction future. In the 2015 film *The Martian*, Matt Damon plays Watney, a fictional botanist turned astronaut who gets stranded on Mars. Watney survives, in part, by constructing a greenhouse, filling it with Martian soil, fertilizing the soil with his poop, and cultivating a whole lot of, as Watney puts it, "all-natural, organic, Martian-grown potatoes."

Still, more ideal than transplanting people to Mars would be preserving the livability of our far superior home planet. "Human spaceflight, being in low Earth orbit, has given us an understanding of the Earth," said Lindgren. As he noted, it was those first photos of our planet taken from space that launched the environmental movement. "It is beautiful, it is gorgeous," he said. "And it is unique."

Lindgren had a hard time keeping his eyes off Earth while on the International Space Station. In the evenings, around 10:00 p.m. Zulu time, the crew would close the aluminum shutters that cover the windows from outside the spacecraft. Sure, it would have been more circadian friendly to do that earlier in the night, he acknowledged, but they loved taking pictures too much. Sometimes Lindgren couldn't pass up a photo op even later in the evening. He would put on his sunglasses, push the dial that slides the shutter open, snap a few photos, and quickly close the shutter again.

From the vantage of low Earth orbit, Lindgren and other astronauts—and thanks to their photographs, the rest of us—have now also witnessed several of humanity's devastating impacts on the Earth, from light pollution to agricultural expansion. Ultimately, it all "impels us to take better care of our home, our planet," Lindgren said. "Whether we reach Mars, whether we establish a colony on Mars, whatever it is we do as we become a multi-planet species, we absolutely have to take care of the Earth."

AS PRECARIOUS AS LIFE WOULD BE ON MARS, HUMANS THREATEN TO transform Earth into a comparably fragile place. We are rapidly decimating soils, forests, biodiversity, and critical reserves of water and energy. As we burn through the last of our fossil fuels, we exacerbate the trouble that the Industrial Revolution started—altering our planet's once relatively moderate climate and filling our skies with more air and light pollution. That changing climate now throws extreme weather at us on the regular. The summer of 2023 was the hottest on record. Wildfires spewed toxic smoke across the US and Canada. That fall, a tropical storm rapidly intensified into a monster hurricane and blindsided the Mexico coast, devastating Acapulco. Climate scientists warn us to expect conditions to grow increasingly dangerous and unpredictable. In some parts of the world, the Earth's surface is already beginning to look almost as inhospitable as it does on Mars.

The consequences of this scorching are staggering. Not the least of which are shortages of food and water. Once again, the knowledge emerging from research on circadian rhythms offers new tools and renewed hope—in this case, for combating climate change and adapting to its effects.

As you may recall, scientists first recognized the circadian clock in plants. Now, plants are one of the first places circadian scientists are looking to apply planet-preserving measures. The pervasiveness of clock control extends to nearly every aspect of a plant's physiology. "Plants, way more so than humans, are tied to circadian rhythms because they can't walk around. They're stuck in the ground," said Steve Kay, a chronobiologist at the University of Southern California. "So plants are exquisitely tuned to their environment." Most genes in *Arabidopsis*, a cabbage-like model plant widely used by scientists, oscillate across the day. Its patterns are comparable to the plants we are more apt to care about, like the ones we eat.

World hunger remains on the rise, despite about half of all habitable land and 70 percent of fresh water on Earth already in use for agriculture. Meanwhile, the global food system—from production to consumption to disposal—is responsible for one third of man-made global greenhouse gas emissions. Conventional agricultural practices also send toxic chemicals seeping onto our landscapes and into our waterways, which have been made more vulnerable to these insults by increasingly regular droughts and floods.

But what if we could produce greater yields while shrinking our agriculture footprint? What if those foods could be healthier? And what if we could use fewer toxic chemicals and less water in the process? Much like how biologists are timing medicines and manipulating mealtimes and lights for people, agricultural and horticultural scientists are attempting to harness and hack the clocks in plants. They are modifying circadian genes, turning on and off lights, and timing inputs to make food and medicinal crops more resistant to extreme conditions, and more bountiful and longer lasting—all the while reducing the use of water, pesticides, and fertilizers. They are even tweaking crop recipes to increase the contents of protein, vitamins, phytochemicals, and cannabinoids.

While many of the efforts underway may one day also support extraterrestrial survival, these are tactics that we can implement here and now and, hopefully, combine with other sustainability efforts to defer the need for a second home. We've manipulated crops while seeking new frontiers before. Take, for example, the advent of agriculture in the Fertile Crescent and other parts of the world around twelve thousand years ago. The transition from foraging to farming meant we could feed more people in one place. It spawned the earliest civilizations. As those populations grew, human settlements spread. People migrating east and west had little trouble growing food to eat. But as people moved north and south, they encountered a problem: their domesticated crops wouldn't grow so easily.

The political geography of our planet may owe a fair bit to circadian clocks. The faster spread of domesticated crops across similar latitudes and climates in Europe and Asia gave those civilizations an edge over populations in Africa and the Americas, where a north-south axis predominates. Over time, by choosing the best growing varieties, the latitudinally challenged farmers slowly and unwittingly selected for crops with traits that helped them thrive in different day lengths and weather conditions. "And all of those strains carry mutations in the circadian clock genes that we discovered in our labs twelve thousand years later," said Kay. "It still makes the hair stand up on my arm."

As farmers moved the tomato plant north through the Americas and Europe, for example, they gradually selected tomatoes with genetic mutations that drove clocks to tick slower. Specifically, the mutations extended the day part of its circadian cycles. We don't know for certain why tomato plants with longer rhythms prosper in the north. One leading theory is that these plants can essentially stay awake longer to take advantage of the additional photosynthesis time during the longer day lengths of summer. Not all plants thrive with slower clocks at higher latitudes, however. And periods tend to shorten along the latitudinal cline for animals.

Scientists also speculate that much of the indirect clock-gene selection over the last few thousand years is linked with changes in the timing of flowering. Plants use their clockworks to detect things like temperature and the start and end of the day, and then incorporate that information into timing longer-term activities like when to flower. A tweak in those cogs and gears could help maximize the plant's yields while minimizing the risk of untimely damage by seasonal frost, heat, or drought in a new locale. We regularly—and more knowingly—exploit this knowledge in compelling "poinsettias to bloom before Christmas, roses to bloom before Valentine's Day, and lilies to bloom before Easter," one scientist told me. That's been fun and all. Still, there are far more imperative reasons to better understand and control plant clocks.

CLIMATE CHANGE IS INTRODUCING AGRICULTURAL CHALLENGES UNprecedented since humanity's past poleward migrations. Warming seasonal temperatures are causing crops to flower too early, ultimately curbing yields. Hotter nights relative to days are again weakening that vital day-night distinction and further confusing clocks. This could mess up how a plant times its photosynthesis and uses its energy stores for growth. But as scientists zero in on the mechanisms involved, they are also finding ways to help crops tolerate higher temperatures. And they are identifying circadian tools that could enable cultivation in new latitudes. Once too cold for certain crops, northern regions are becoming more hospitable and could take over agricultural duties from places growing too warm for the job. My home state of Washington is becoming a US hotbed for growing wine grapes—a job previously reserved for

California. But relocated plants are forced to contend with unfamiliar periods of daylight. With each northern move, rather than wait through the lengthy domestication process, scientists could genetically alter the period of crop clocks to fit local day lengths and, therefore, to tell the crop to flower at an appropriate time. If ultimately necessary, the same edits may end up suitable for adaptation to the long Mars sol.

Taking the ten-thousand-foot view, scientists also warn of snowballing circadian consequences in our forests. New research, published as a preprint in March 2024, hints that a hotter climate could thwart the ability of some trees to track time, potentially lowering their chances of survival—or at least reducing their capacity to sequester carbon.

As we face a changing climate, dwindling pollinator populations, and a growing human population, humanity might need to go to even greater extremes. We might need to, as Matt Damon's character put it, "science the shit" out of crops. Circadian scientists are investigating strategies to engineer self-pollinating sunflowers to sidestep pollinator losses. They are looking at how to manipulate plant clocks to extend the hours of the day and the months of the year that a crop can grow. Unexpected outcomes and trade-offs may be inevitable, however, especially as we seek competing goals—like cultivating a grain that grows around the clock but is also resistant to drought.

From Afghanistan to Italy to the Central Valley of California, the immense impact of drought is clear. It's evident from astronauts' photographs from space. More than three quarters of the world could face drought by 2050. Modern crops might not be ready to cope. While their wild ancestors acquired survival traits over millennia, breeding crops for more marketable attributes essentially overwrote many of those traits from the plant's genome.

In addressing this challenge, one focus is on the stomata, tiny pores in plant leaves and stems that open and close in an approximately twenty-four-hour rhythm. A quick refresher on photosynthesis via that old high school textbook: Plants capture sunlight and use that energy to turn carbon dioxide and water into oxygen and energy they store as sugar. All that carbon dioxide and oxygen enters and exits through stomata. When the stomata are more open to support this cycle, the plant loses more water. When the stomata close, the plant retains water, yet photosynthesis slows down or even stops. It's a balance game—regulated by the clock. By manipulating the time of day at which stomata open via modifications to clock genes, scientists could reduce how much water crops need. More simply and cheaply, farmers can also adjust when they water crops—made even easier with the rise of automated irrigation systems.

Climate change is also fanning the spread of insects, pathogens, and weeds. Pests already destroy up to 40 percent of global crops every year. But they don't randomly inflict their damage. A certain pest is more likely to attack a certain plant at certain times of day. And that plant has learned that pest's schedule. In much the same way as a plant creates and releases chemicals to attract pollinators ahead of their typical visiting times, it will also build up levels of less benign or palatable chemicals in anticipation of being preyed upon. One study found that a plant and an insect pest in the same light-dark cycle will reach a stalemate. But if you artificially place the insect on a light-dark cycle shifted twelve hours so that it's opposite from the plant, then the insect will devastate the plant. The plant is caught off guard without its weapons on hand. These circadian-regulated defenses, too, could be fine-tuned and harnessed to protect valuable crops.

A farmer's arsenal typically includes a range of pesticides. As with pharmaceutical drugs, we now know that a standard dose will have different effects depending on when it is applied to a field. And as with prescribing doctors, few farmers know about these time-of-day effects. Even fewer leverage them to keep unwelcome squatters out of their fields or orchards—troublemakers like the invasive spotted wing drosophila.

The insect is a relative of those fruit flies widely used in labs, and my friend Per. It is also a major fruit crop pest, decimating cherries, strawberries, blueberries, and raspberries around the world. In California, which produces most of the nation's fruits and nuts, the pest is rapidly developing resistance to insecticides. Joanna Chiu, a chronobiologist and entomologist at the University of California, Davis, investigated whether adjusting the time of day that an insecticide is applied might salvage that dwindling effectiveness. Her team sprayed the spotted wing drosophila with insecticides in the lab to mimic field treatments. Because many detoxifying systems in the insect's body are regulated in a circadian manner, spraying the insects at the correct hour could render them less able to tolerate the insult. More will die. And that's just what she and her team found in a study simulating summer and winter days in Watsonville, California, where the spotted wing drosophila was first detected. The costly scourge was most susceptible to the popular insecticide malathion in the early morning.

The same principle holds for targeting weeds. In this case, the goal is to let down the guard of a nuisance plant. Glyphosate, the main ingredient in Monsanto's signature Roundup herbicide, has been deemed the most widely and heavily applied weed killer in the history of chemical agriculture. It's hugely popular among conventional farmers who spray it on fruits, vegetables, and nuts, as well as on global staples, such as corn and soybeans, that have been genetically engineered to resist glyphosate. Researchers have discovered that, at certain times of day, farmers need to spray far less of the herbicide to achieve the target effect. Reducing application of Roundup could mean less cost to the farmer and less exposure to a chemical that some studies have linked with health problems, including certain cancers. Of course, Monsanto might not be thrilled with a drop in sales.

Circadian science is also pointing to chemical-free and biotech-free alternatives for controlling plants. In his office at the University of Cambridge, plant scientist Alex Webb showed me a stack of yellowed scientific papers: "On the Artificial Production of Rhythm in Plants," "The Effect of Light on the Transpiration of Leaves," and "Observations on Stomata." All were authored in the 1800s by Francis Darwin, the son of Charles, while he worked in the same plant science department.

The papers stress the importance of light in signaling the time of day for a plant. Unlike us, every cell in a plant is capable of sensing light.

They have photoreceptors that detect colors beyond our range of perception, absorbing everything from ultraviolet to infrared. Webb told me how Francis Darwin and his dad grew very interested in the role of red and blue light, especially in the opening and closing of the stomata. And they got creative. The Darwins filled up a massive glass tank with water and placed it in front of a window so the light would shine through. To test the effects of blue light on plants, they added copper sulfate to the tank. To test red light, they added wine.

Color-tunable lighting has progressed since Darwin's day. Novel LED technology and a greater understanding of light's influence on biology have revolutionized its applications for people and plants alike. Wine grapes happen to be a beneficiary. Ultraviolet light shows promise in protecting crops against pathogens such as powdery mildew, a disease notorious for decimating grapes, strawberries, tomatoes, and cannabis. The fungi that cause powdery mildew, which appears as a whitish-gray powder coating leaves or fruit, is developing resistance to various fungicides. So scientists have enlisted a new tool: GPS-driven robots fitted with ultraviolet lamps that shine light at night on fields and inside greenhouses. They are taking advantage of the fact that powdery mildew evolved to turn on mechanisms that repair DNA damage from the sun's ultraviolet light only during the day. At night, in the absence of blue light, the fungus puts its guard down and becomes vulnerable. Trials so far show that precisely timed doses of ultraviolet light are sometimes superior, other times inferior, to available fungicides in effectively killing the pest while not damaging the plant.

AS WITH CIRCADIAN MEDICINE, THE IMPRACTICALITY OF APPLYING time-of-day schemes can be a roadblock to its use on the farm. "We do things on the farm because that's when we're awake and can see, not because it's the right time," said Webb. But technological advances are now allowing farmers to bypass the human clock. Cheap sensors and robots enable them to collect data and tend to their fields at all hours of day and night. You might imagine a drone flying over a field, running an analytical program that determines the circadian phase of a specific crop in a particular location under specific weather and soil conditions. A farmer could feed that data into a computer that tells them when to apply a pesticide, fertilizer, water, or LED lights. And, if it is not a convenient time for the farmer to deploy an agrochemical, for example, they could tell a robot to do it. "That could save enormous amounts of money," another scientist told me. "It could reduce substantially the amount of agrochemical runoff. It could reduce the amount of product that remains within something sold to consumers."

Growing crops indoors eliminates variables and allows for even more automation while potentially shrinking agriculture's carbon and literal footprints. Farmers can produce food closer to where it will be consumed, like a city center. They can do so within shipping containers, vacant buildings, even deserted subterranean spaces fit with vertically stacked layers of beds. And no longer is a crop's endurance at the whim of nature. With weather conditions increasingly variable and frequently devastating traditional outdoor farms, the concept is growing in appeal. However, indoor agriculture will never be the sole solution. It is not very practical for major crops like wheat or potatoes. Plus, it hogs energy—unless you live in a region with abundant sunshine or another resource on hand to power the operation. Vertical farms are on the rise in Brazil, where hydroelectric dams provide relatively inexpensive renewable energy. And Iceland, where the period of daylight shrinks to just four hours in the winter, has been tapping into plentiful geothermal energy to grow crops indoors for at least a century.

Webb directed me to one London-based company, Vertical Future, which he has worked with to fine-tune various clock-tunable variables: light, temperature, water, and nutrients. I met Jim Stevens, an investment analyst turned plant scientist, inside the company's redbrick building, not far from the Greenwich Observatory. He and I donned white lab coats, blue hairnets, and booties before venturing inside his test "fields." Vertical Future uses no insecticides, herbicides, or fungicides, so we needed to be extra careful to avoid tracking in contaminants. Stevens then led me up and down several rows of crops. Each was stacked ten beds high. Above each layer, arrays of multicolored LED lights combined to give a strange purple-pink glow to the whole space.

I was surrounded by hundreds of edible plants, from herbs such as mint, sage, and coriander, to protein-rich amaranth and a variety of peppers, tomatoes, and leafy greens. I sampled a selection. The sorrel, a popular green in restaurants, tasted citrusy. The Basket of Fire pepper was as you might guess: hot. Also within these stacked beds were plants grown for compounds used in cancer drugs and vaccines. Every one of these species, Stevens explained, has its preferences for light spectrums and day lengths. And they don't always match what nature offers outside the brick walls. This is one case where artificial light might outshine the gold standard of the sun. In consultation with Webb, Stevens is investigating each plant's ideal light mixes and application times. Just like for people, the color, intensity, timing, duration, even the direction at which the light hits the plant might matter. The time it's harvested could affect the quality of the product, too.

Vertical Future has developed a high-tech system that automatically tweaks lights and other inputs around the twenty-four-hour local clock, using knowledge of the plant clocks. It sells these systems along with vertical farming equipment and light recipes. Clients can subscribe to get recipe updates as the science progresses.

Before I left, Stevens gave me a paper bag full of freshly harvested spinach from the indoor farm. It spent the night in my Airbnb fridge. Sadly, when I opened the fridge, seeking to snack on the spinach the next day, all the leaves were limp.

THE CLOCK CAN KEEP TICKING FOR A PLANT LONG AFTER HARVEST. But we humans frequently cripple that clockwork and speed its demise.

You've probably witnessed this phenomenon without realizing it, especially if you've ever bought a head of iceberg lettuce. Whether it is grown in an outdoor field or in an indoor farm, typical store-bound lettuce goes through daily light and dark cycles. It is then harvested and shipped off. If it lands in a twenty-four-hour supermarket, a head of lettuce will likely lay under lights all day and all night. But after the lettuce is brought home and put inside a refrigerator, the only light that will reach its leaves is the occasional blast every time the fridge door opens.

Like us, when produce is kept under constant environmental conditions, whether twenty-four-hour light or dark, its circadian rhythms will dampen. Also like in us, that will cause its tissues and defenses to weaken. It will start to lose its nutrient content, prematurely wilt, and become more vulnerable to pests and disease. And it may develop those black or brown spots you've probably seen tainting a head of lettuce. Maybe you, too, have given up and thrown it out—as I did with my floppy spinach. Sorry, Jim.

I wish that was the last time I failed to consume produce before it was too late. While my rotten fruits and vegetables frequently invite visits from flies who stir memories of Per, I'd still be happy to break the habit. More than a third of all food produced worldwide goes to waste. However, the strategic use of light now shows promise in prolonging shelf life and letting lettuce live to see another day. Research has shown that light can indeed re-entrain the rhythms of produce, including blueberries, carrots, and sweet potatoes. That's the idea behind the Harvest-Fresh technology, marketed by the appliance company Beko. Red, green, and blue LED lights in produce drawers simulate the day-night cycle to keep clocks ticking for harvested fruit and vegetables—and hopefully helping them stay fresh longer, maintain more of their nutritional value, and better evade pest damage.

This novel science might also inspire methods to preserve freshness without the use of energy-intensive refrigerators at all. Rather than shipping crops from California to New York in refrigerated trucks, we could cycle energy-efficient LEDs. In one study, the appearance, taste, and nutritional value of kale and cabbage remained higher over time when stored under light-dark cycles designed to maintain biological rhythms compared with plants without the cycled lighting. The longevity of the greens was on par with refrigerated produce. Pulses of light might be enough to set or control a plant's clock—similar to the novel findings for calibrating our own rhythms. Solving food insecurity will require the more equitable distribution of food products. This might be one tool that could make that more feasible.

Another global health challenge has been the precipitous decline in the nutritional value of crops, largely resulting from depleted soils and the widespread use of agrochemicals. While we work to improve the health of our soils—potentially through tweaking microbe and plant clocks—we can also mind the clock to take greater advantage of the vitamins and other healthy compounds still available. The nutrient content of a crop may vary based on the time of day it is harvested. It may even vary based on the time of day we consume it—if the crop's clock is still ticking. What's more, how our bodies utilize those nutrients may vary depending on our own circadian phase.

Circadian science may afford further opportunities to support human health and environmental sustainability. If we knew when during the day a particular bacterium is most susceptible to a particular antibiotic, for example, then we may be able to lower the doses prescribed to human patients, administered to livestock, and spread over crop fields. Not only would that save the money and lessen the microbiome and circadian disruptions associated with antibiotic use, it could also mean fewer traces of antibiotics lingering in our environment and in our food—major contributors to the pressing public health threat of antibiotic resistance. Scientists are also looking to coax clocks in cyanobacteria and sunflowers—among other plants, fungi, and bacteria—to create more efficient living factories for pharmaceuticals, cosmetics, bioplastics, and biofuels. Soon, people with diabetes could benefit from greater availability and affordability of insulin. And more prolific production of transport biofuel might one day help us both sustain our planet and venture to others.

Martha Merrow, the chronobiologist in Munich, also sees a market for manipulating yeast clocks: better beer fermentation. That may be less critical for saving our home planet, yet it may be welcome for drowning our sorrows should you or I need to relocate to a starker, dustier one—or descend deep into a bunker.

THE TITAN II MISSILE THAT ONCE OCCUPIED GT HILL'S BUNKER WAS part of a Cold War arsenal that inspired the creation of the Doomsday Clock, a symbol representing the threats to humanity from unchecked technological advances. The clock stood at one hundred seconds to midnight when I stayed underground in late 2021. Scientists have since moved the hands of the clock farther forward to ninety seconds, in part due to mounting dangers posed by Russia and its nuclear arsenal. They also cite climate change and biological risks, such as COVID-19.

Is it time to crawl back into that bunker? Hill had told me his dream of an underground garden, and applying circadian science could help it flourish. Still, even with fresh veggies and perhaps the company of loved ones the next time, the idea appeals even less than moving into a Martian colony with fellow Earth refugees. Greater international cooperation to combat climate change and fend off nuclear war sounds far better. The predicament in *The Martian* tightened friendly fictional ties between the US and China. Could we follow Hollywood's lead? Kjell Lindgren, the real astronaut, suggested so—at least in the context of space. "I think that exploration is incredibly important," he told me. "By accepting challenges, by reaching for things beyond our grasp, it forces us to work together to cooperate as an international partnership. It forces us to innovate."

We could pursue innovations that simultaneously strengthen plant and planetary defenses. How about growing crops under solar panels? Scientists have found that we might maximally harvest food and energy with a system that allocates specific wavelengths of sunlight to each activity. The red part of the spectrum could feed the plants, for example, while the blue part feeds the production of solar energy. As we've learned, we should be taking advantage of all of nature's wavelengths.

There is hope for us, individually and collectively. You and I can start with our personal circadian hygiene. As cities, states, and nations, we can begin to brighten the designs of our indoor environments. We can turn down our lights at night—except maybe inside the greenhouses where we grow sustainable food, medicine, and fuel. We can reconsider how we wind our clocks forward and back, and how we organize our work and school days. And, as a united world, we can think more critically about those arbitrary lines we draw on the map. We've ignored circadian rhythms for too long, to our peril. Not only have we lost a fundamental connection with the rhythms of nature—the cycles of the seasons, the moon, the sun—but the quality and quantity of our years on this beautiful blue planet, and those of future generations, is at stake. It's time to rekindle those relationships. It's time to reset and recover our inner clocks.

Acknowledgments

The Inner Clock would have never seen the light of day, nor the LED light of bookstores, without the collaborative efforts of many people. My skilled agent Suzanne Gluck at WME acted as a well-tuned circadian clock, setting the book for success at the start and supporting its progress with timely guidance. I'm deeply thankful to her for believing in me and championing this project. I also had the great fortune to work with Courtney Young at Riverhead Books, who wielded her editorial genius at all levels—from patiently coaching me to shape a deluge of material into a coherent whole to adeptly managing all the moving parts of the project. Her compassion and humor consistently buoyed me. More talented people at Riverhead contributed to pulling this book together and putting it into the hands of readers: Catalina Trigo, Caitlin Noonan, Brian Borchard, Mike Brown, Denise Boyd, Angelica Krahn, Daniel Lagin, Claire Vaccaro, Ashley Garland, Kitanna Hiromasa, Nora Alice Demick, Viviann Do, Grace Han, Geoff Kloske, and Jynne Dilling Martin. I owe further gratitudes across the pond to Matilda Forbes Watson at WME and to the team at Bloomsbury in the UK, who brilliantly adapted and promoted this book: Jasmine Horsey, Molly McCarthy,

David Mann, Mireille Harper, Grace Nzita-Kiki, Youssef Khaireddine, Fabrice Wilmann, and Thea Hirsi.

Tackling the topics covered in these pages was a fantastic challenge. Hundreds of scientists, engineers, architects, athletes, teachers, students, patients, health-care workers, military service members, and others graciously sat with me, chatted with me, and responded to email after email—and then had the endurance to respond to "just one more question." Many of the people who were so generous with their time and knowledge don't appear on these pages, but all our conversations profoundly informed my understanding of the science and its implications.

A few special thanks are in order: GT Hill went above and beyond to accommodate my highly unusual requests at the Titan Ranch, even randomizing his own sleep-and-wake schedule to ensure my experiment wasn't derailed. Lora Lee Wicks and her family welcomed me to camp in their sea of dancing sunflowers. Christopher Jung—along with Klint, Sean, Steve, Larry, and Scout—protected me from getting swallowed by a river or trampled by a moose, while sharing with me the light of day *and night*. Alicia Rice spent intimate hours with my saliva to decode my DLMO. Benjamin Smarr, Till Roenneberg, Elizabeth Klerman, Robert Soler, Erik Page, David Basken, Sofia Axelrod, Christina Friis Blach Petersen and the team at LYS, and Achim Kramer and the team at Body-Clock contributed time and tools to further decipher the ticking of my inner clocks and the light memos they translate. Smarr, along with Ethan Buhr and Olivia Walch, dug deep into my data and created colorful images of that symphony of timekeepers. William Schwartz and Vincent Cassone shared enthralling history lessons. John Hogenesch, Richard Lang, Horacio de la Iglesia, Jay Neitz, Russell Van Gelder, Carrie Partch, and Priya Crosby hosted multiple enlightening and inspiring visits. So many others took exceptional time to tour me through their labs, their homes, their towns, and other unique places of interest such as a historic synagogue and a Coast Guard icebreaker: Lisa Heschong, Marty Brennan, Camilla Kring, Toine Schoutens, Alex Webb, John Mardaljevic, Rabbi Shalom Morris, Jens Christoffersen, William Woityra, Orie Shafer, Debra Skene, Rae Silver, and Susan Golden, to name a few.

During my research, I drew on the scholarship of many excellent books including *Internal Time* by Till Roeneberg, *Why We Sleep* by Matthew Walker, *Life Time* by Russell Foster, *The Circadian Code* by Satchin Panda, *When* by Daniel Pink, and *Time, Love, Memory* by Jonathan Weiner. This book also builds on articles I wrote for *Undark Magazine* and *Nature.* I owe thanks to those editorial teams—especially Tom Zeller, Jr., John Montorio, and Brendan Maher—for commissioning and deftly editing the stories, portions of which appear in revised form in chapters ten and twelve.

I am indebted to Deborah Blum, Ashley Smart, and the entire team at the MIT-Knight Science Journalism Program, whose fellowship gave me the financial support, expert guidance, and time to put together the proposal for this book. Another enormous thank-you to Doron Weber and the Alfred P. Sloan Foundation for their generous funding and for their important efforts to promote the public understanding of science.

I wish I could say I maintained robust circadian rhythms throughout the process of researching and writing *The Inner Clock.* I wish I could say that I never once missed a bedtime or a mealtime or morning light, and that I never once sipped wine or walked through an intensely lit grocery store after 8:00 p.m. Had I claimed so, I'm sure my meticulous and hilarious fact checker Emily Krieger would have called me out. In addition to making astute catches, she elicited welcome and well-timed laughs that saved my sanity. (Any remaining mistakes are on me.) An additional thank-you to Carrie Partch for her thorough scientific review of the manuscript. Mike Goldstein and Erica Westly also provided insightful early readings of chapters, and my beloved Bookies added wise direction as I navigated the final stretch to this book's completion.

I am lucky have a network of talented, inspiring science writing pals in Seattle and across the country. Carina Storrs, Roberta Kwok, Bryn Nelson, and Michael Bradbury offered particularly keen advice and kindness throughout this journey. Of course, this journey really began years ago. Countless teachers, professors, and mentors have fostered my scientific curiosity and shaped my writing career, including Doug Kimball, Steve McKelvey, Dan Fagin, Robin Lloyd, Tom Zeller Jr., and more colleagues and editors from my green years on the Green Team at *The Huffington Post* and beyond. My father Clint Peeples, a former high school science teacher, belongs at the top of this influencer list. Spending time with him hiking up mountains, jumping in ocean waves, and staring at starry skies instilled in me a love of nature and its rhythms. He shares that top spot with my late mother Judy Peeples, who was my first and foremost writing instructor and cheerleader. I very much miss her sharp red pen, clever ideas, and hugs.

Finally, even while I siloed myself away in a bunker and apartment to research and write, I consistently felt the love of my community of peeps. A heartfelt thanks to Linda Yoo and Don Haig for the encouragement—and for the daylight and view. And huge hugs of gratitude to my dad, my brother, Gregg Peeples, and all my friends. This book would never have been possible without your support and enthusiasm. I appreciate all the times you indulged my tangents as I injected circadian science into just about every topic of conversation. Thank you also for giving me the time and space to obsess, and the liberty to pester you about turning down the lights at night. Now, go blow out (or click off) your candles and go to bed.

Notes

Included here are extensive but far from comprehensive citations. For more complete notes, please visit lynnepeeples.com/the-inner-clock.

INTRODUCTION

x **This can sabotage sleep:** A small sample of the studies that draw these connections: Junyan Duan, Elyse Noelani Greenberg, Satya Swaroop Karri, and Bogi Andersen, "The Circadian Clock and Diseases of the Skin," *FEBS Letters* 595, no. 19 (October 2021): 2413–36, doi.org/10.1002/1873-3468.14192; Sarah L. Chellappa, Nina Vujovic, Jonathan S. Williams, and Frank A. J. L. Scheer, "Impact of Circadian Disruption on Cardiovascular Function and Disease," *Trends in Endocrinology and Metabolism* 30, no. 10 (October 2019): 767–79, doi.org/10.1016/j.tem.2019.07.008; Robin M. Voigt, Christopher B. Forsyth, and Ali Keshavarzian, "Circadian Rhythms: A Regulator of Gastrointestinal Health and Dysfunction," *Expert Review of Gastroenterology & Hepatology* 13, no. 5 (March 2019): 411–24, doi.org/10.1080/17474124.2019.1595588.

x **The consequences of circadian disruption:** Peng Zhang et al., "Environmental Perturbation of the Circadian Clock during Pregnancy Leads to Transgenerational Mood Disorder-like Behaviors in Mice," *Scientific Reports* 7, no. 1 (October 2017): 12641, doi.org/10.1038/s41598-017-13067-y.

xii **How about the 99 percent:** Fabio Falchi et al., "The New World Atlas of Artificial Night Sky Brightness," *Science Advances* 2, no. 6 (June 2016): e1600377, doi.org/10.1126/sciadv.1600377.

xii **composition of microbial squatters:** Yueliang Zhang et al., "The Microbiome Stabilizes Circadian Rhythms in the Gut," *Proceedings of the National Academy of Sciences of the United States of America* 120, no. 5 (January 2023), e2217532120, doi.org/10.1073/pnas.2217532120.

xii **restricting meals to daylight:** Emily N. C. Manoogian et al., "Time-Restricted Eating for the Prevention and Management of Metabolic Diseases," *Endocrine Reviews* 43, no. 2 (April 2022): 405–36, doi.org/10.1210/endrev/bnab027.

xii **Even adjusting the time chemotherapy:** Pasquale F. Innominato et al., "The Future of Precise Cancer Chronotherapeutics," *Lancet Oncology* 23, no. 6 (June 2022): e242, doi.org/10.1016/S1470-2045(22)00188-7.

xiii **Our immune system is under:** Christoph Scheiermann, Yuya Kunisaki, and Paul S. Frenette, "Circadian Control of the Immune System," *Nature Reviews Immunology* 13, no. 3 (March 2013): 190–98, doi.org/10.1038/nri3386.

xvi **Some moves are surprisingly simple:** Jennifer Hahn-Holbrook et al., "Human Milk as 'Chrononutrition': Implications for Child Health and Development," *Pediatric Research* 85, no. 7 (June 2019): 936–42, doi.org/10.1038/s41390-019-0368-x.

CHAPTER 1. LOSING TIME

7 **color least likely:** "Module 2. How Shift Work and Long Work Hours Increase Health and Safety Risks," in *NIOSH Training for Nurses on Shift Work and Long Work Hours*, National Institute for Occupational Safety and Health, 2020, accessed April 2, 2020, cdc.gov/niosh/work-hour-training-for-nurses/longhours/mod2/20.html.

11 **all three mutations:** Ronald J. Konopka and Seymour Benzer, "Clock Mutants of *Drosophila melanogaster*," *Proceedings of the National Academy of Sciences of the United States of America* 68, no. 9 (September 1971): 2112–16, doi.org/10.1073/pnas.68.9.2112.

11 **"stumbled straight into the center":** Jonathan Weiner, *Time, Love, Memory: A Great Biologist and His Quest for the Origins of Behavior* (New York: Alfred A. Knopf, 1999).

13 **On July 16, 1962:** Michel Siffre, *Beyond Time* (London: Chatto and Windus, 1965).

14 **"My co-workers and I":** Jürgen Aschoff, "Circadian Rhythms in Man," *Science* 148, no. 3676 (1965): 1427–32, doi.org/10.1126/science.148.3676.1427.

14 **Aschoff's team asked participants:** The researchers made the bunker apartments as comfortable as possible. There was a bedroom with a cushioned chair, table, and desk. There was a shower and a small kitchen. Artificial lights could be turned on and off. This would become a source of criticism among scientists who feared participants' control of the lights altered outcomes.

15 **Charles Czeisler, a sleep and circadian researcher:** Kenneth P. Wright Jr. et al., "Intrinsic Near-24-h Pacemaker Period Determines Limits of Circadian Entrainment to a Weak Synchronizer in Humans," *Proceedings of the National Academy of Sciences of the United States of America* 98, no. 24 (November 2001): 14027–32, doi.org/10.1073/pnas.201530198.

15 **widely considered the most:** J. F. Duffy et al., "Sex Difference in the Near-24-Hour Intrinsic Period of the Human Circadian Timing System," *Proceedings of the National Academy of Sciences of the United States of America* 108, Supplement 3, (September 2011), doi:10.1073/pnas.1010666108.

CHAPTER 2. WHAT MAKES YOU TICK?

This chapter leans largely on history lessons from several circadian scientists, including Vincent Cassone, a biologist at the University of Kentucky, and William Schwartz, a neuroscientist at The University of Texas at Austin. Several sources are available for a deeper dig into the history, such as: William J. Schwartz and Serge Daan, "Origins: A Brief Account of the Ancestry of Circadian Biology," in *Biological Timekeeping: Clocks, Rhythms and Behaviour* (New Delhi: Springer, India, 2017), 3–22.

28 **I doubt anyone would pay:** Filipa Rijo-Ferreira and Joseph S. Takahashi, "Sleeping Sickness: A Tale of Two Clocks," *Frontiers in Cellular and Infection Microbiology* 10 (October 2020): 525097, doi.org/10.3389/fcimb.2020.525097.

29 **Our strength generally maxes out:** Collin M. Douglas, Stuart J. Hesketh, and Karyn A. Esser, "Time of Day and Muscle Strength: A Circadian Output?," *Physiology* 36, no. 1 (January 2021): 44–51, doi.org/10.1152/physiol.00030.2020.

29 **Circadian rhythms could serve:** Henrik Oster, Erik Maronde, and Urs Albrecht, "The Circadian Clock as a Molecular Calendar," *Chronobiology International* 19, no. 3 (May 2002): 507–16, doi.org/10.1081/CBI-120004210.

30 **It's a strategy reminiscent of:** William Rowan, "Light and Seasonal Reproduction in Animals," *Biological Reviews of the Cambridge Philosophical Society* 13, no. 4 (October 1938): 374–401, doi:10.1111/j.1469-185x.1938.tb00523.x.

31 **That's when Jean-Jacques d'Ortous:** Jean-Jacques d'Ortous de Mairan, "Observation Botanique," *Histoire de l'Academie Royale des Sciences* (January 1729): 35. Translated from the French.

33 **paper published in 1898:** John H. Schaffner, "Observations on the Nutation of *Helianthus annuus*," *Botanical Gazette* 25, no. 6 (June 1898): 395–403, jstor.org/stable/2464526.

34 **Harmer and her team concluded:** Hagop S. Atamian et al., "Circadian Regulation of Sunflower Heliotropism, Floral Orientation, and Pollinator Visits," *Science* 353, no. 6299 (August 2016): 587–90, doi.org/10.1126/science.aaf9793.

37 **Among the first brains:** Junko Nishiitsutsuji-Uwo and Colin S. Pittendrigh, "Central Nervous System Control of Circadian Rhythmicity in the Cockroach: III. The Optic Lobes, Locus of the Driving Oscillation?," *Journal of Comparative Physiology* 58, no. 1 (March 1968): 14–46, doi.org/10.1007/BF00302434; Friedrich K. Stephan and Irving Zucker, "Circadian Rhythms in Drinking Behavior and Locomotor Activity of Rats Are Eliminated by Hypothalamic Lesions," *Proceedings of the National Academy of Sciences of the United States of America* 69, no. 6 (June 1972): 1583–86, doi.org/10.1073/pnas.69.6.1583; and Curt P. Richter, "'Dark-Active' Rat Transformed into 'Light-Active' Rat by Destruction of 24-Hr Clock: Function of 24-Hr Clock and Synchronizers," *Proceedings of the National Academy of Sciences of the United States of America* 75, no. 12 (December 1978): 6276–80, doi.org/10.1073/pnas.75.12.6276.

38 **A series of complex:** Michael H. Hastings, Elizabeth S. Maywood, and Marco Brancaccio, "The Mammalian Circadian Timing System and the Suprachiasmatic Nucleus as Its Pacemaker," *Biology* 8, no. 1 (March 2019): 13, doi.org/10.3390/biology 8010013.

39 **This push-and-pull process:** While the dogma remains in the field that circadian rhythms are driven by so-called transcription-translation feedback loops, some scientists suggest that the concept might not be sufficient or even necessary to explain daily rhythms in our physiology. A few studies show that rhythmic protein activity alone can generate daily rhythms. Redundant processes may reinforce each other to increase circadian robustness. As this science continues to evolve, we'll focus primarily on the output of this "black box." Alessandra Stangherlin, Estere Seinkmane, and John S. O'Neill, "Understanding Circadian Regulation of Mammalian Cell Function, Protein Homeostasis, and Metabolism," *Current Opinion in Systems Biology* 28 (December 2021): 100391, doi.org/10.1016/j.coisb.2021.100391.

39 **still in the womb:** Keenan Bates and Erik D. Herzog, "Maternal-Fetal Circadian Communication during Pregnancy," *Frontiers in Endocrinology* 11 (April 2020): 198, doi.org/10.3389/fendo.2020.00198.

39 **patients with schizophrenia:** Madeline R.Scott, Wei Zong, Kyle D. Ketchesin,

Marianne L. Seney, George C. Tseng, Bokai Zhu, and Colleen A. McClung, "Twelve-Hour Rhythms in Transcript Expression within the Human Dorsolateral Prefrontal Cortex Are Altered in Schizophrenia," *PLoS Biology* 21, no. 1 (January 2023): e3001688, doi:10.1371/journal.pbio.3001688.

40 **Scientists disagree on whether:** Many scientists suggest that cyanobacteria or something similar likely evolved the first circadian clock. However, other lifeforms may have still independently evolved clocks that converged on the same basic time-keeping properties.

40 **"If life exists on other":** Hans P. A. Van Dongen et al., "A Circadian Biosignature in the Labeled Release Data from Mars?," in "Astrobiology and Planetary Missions," ed. Richard B. Hoover, Gilbert V. Levin, Alexei Y. Rozanov, and G. Randall Gladstone, *Proceedings of SPIE* 5906 (September 2005): 107–16.

41 **By adding three more proteins:** Archana G. Chavan et al., "Reconstitution of an Intact Clock Reveals Mechanisms of Circadian Timekeeping," *Science* 374, no. 6564 (October 2021): eabd4453, doi.org/10.1126/science.abd4453.

43 **His lab followed up:** Jérôme Wuarin et al., "The Role of the Transcriptional Activator Protein DBP in Circadian Liver Gene Expression," in "Transcriptional Regulation in Cell Differentiation and Development," ed. Peter Rigby, Robb Krumlauf, and Frank Grosveld, supplement, *Journal of Cell Science*, no. Supplement 16 (January 1992): 123–27; Jérôme Wuarin and Ueli Schibler, "Expression of the Liver-Enriched Transcriptional Activator Protein DBP Follows a Stringent Circadian Rhythm," *Cell* 63, no. 6 (1990): 1257–66, doi:10.1016/0092-8674(90)90421-a.

43 **Scientists, including Michael Menaker:** S. Yamazaki et al., "Resetting Central and Peripheral Circadian Oscillators in Transgenic Rats," *Science* 288, no. 5466 (April 2000): 682–85, doi:10.1126/science.288.5466.682.

43 **approximately ten trillion cells:** Our bodies contain about thirty trillion cells. Approximately twenty trillion of these are red blood cells, which lack a nucleus and therefore circadian clock machinery. The activity of red blood cells does still follow a daily rhythm, however, presumably as a downstream effect of clocks in other cells. And then there are the trillions of bacteria cells we also harbor.

43 **As the clock genes:** Clock genes can do more than just inform circadian timing, which can complicate the interpretation of research studies that link circadian clock function to various health conditions. Ray Zhang et al., "A Circadian Gene Expression Atlas in Mammals: Implications for Biology and Medicine," *Proceedings of the National Academy of Sciences of the United States of America* 111, no. 45 (2014): 16219–24, doi.org/10.1073/pnas.1408886111; and Kimberly H. Cox and Joseph S. Takahashi, "Circadian Clock Genes and the Transcriptional Architecture of the Clock Mechanism," *Journal of Molecular Endocrinology* 63, no. 4 (November 2019): R93–102, doi:10.1530/JME-19-0153.

43 **Each of the twenty thousand:** David K. Welsh, Joseph S. Takahashi, and Steve A. Kay, "Suprachiasmatic Nucleus: Cell Autonomy and Network Properties," *Annual Review of Physiology* 72 (March 2010): 551–77, doi.org/10.1146/annurev-physiol-021909-135919.

CHAPTER 3. POWER HOURS

48 **From that data, Ptacek:** C. R. Jones et al., "Familial Advanced Sleep-Phase Syndrome: A Short-Period Circadian Rhythm Variant in Humans," *Nature Medicine* 5, no. 9 (September 1999): 1062–65, doi.org/10.1038/12502.

51 **sleep is of a higher quality:** Andrew J. K. Phillips et al., "Irregular Sleep/Wake Patterns Are Associated with Poorer Academic Performance and Delayed Circadian

and Sleep/Wake Timing," *Scientific Reports* 7, no. 1 (June 2017): 3216, doi.org/10.1038/s41598-017-03171-4; and Matthew D. Weaver et al., "Adverse Impact of Polyphasic Sleep Patterns in Humans: Report of the National Sleep Foundation Sleep Timing and Variability Consensus Panel," *Sleep Health* 7, no. 3 (June 2021): 293–302, doi.org/10.1016/j.sleh.2021.02.009.

53 **Researchers found that members:** David R. Samson et al., "Chronotype Variation Drives Night-Time Sentinel-like Behaviour in Hunter–Gatherers," *Proceedings of the Royal Society B: Biological Sciences* 284, no. 1858 (July 2017): 20170967, doi.org/10.1098/rspb.2017.0967.

54 **The modern chronotype curve has:** Several papers coauthored by Till Roenneberg describe the epidemiology of our chronotypes and the implications of this diversity, including Till Roenneberg et al., "Epidemiology of the Human Circadian Clock," *Sleep Medicine Reviews* 11, no. 6 (December 2007): 429–38, doi.org/10.1016/j.smrv.2007.07.005; and Till Roenneberg, Eva C. Winnebeck, and Elizabeth B. Klerman, "Daylight Saving Time and Artificial Time Zones—a Battle between Biological and Social Times," *Frontiers in Physiology* 10 (August 2019): 944, doi.org/10.3389/fphys.2019.00944.

56 **Just a few days:** Ellen R. Stothard et al., "Circadian Entrainment to the Natural Light-Dark Cycle across Seasons and the Weekend," *Current Biology* 27, no. 4 (February 2017): 508–13, doi.org/10.1016/j.cub.2016.12.041.

57 **In our later years:** Russell Foster, *Life Time: The New Science of the Body Clock, and How It Can Revolutionize Your Sleep and Health* (London: Penguin Life, 2022), 182.

57 **Roenneberg and his colleagues coined:** Marc Wittmann, Jenny Dinich, Martha Merrow, and Till Roenneberg, "Social Jetlag: Misalignment of Biological and Social Time," *Chronobiology International* 23, no. 1–2 (2006): 497–509.

58 **The consequences accumulate:** Till Roenneberg, "How Can Social Jetlag Affect Health?" *Nature Reviews Endocrinology* 19, no. 7 (July 2023): 383–84, doi.org/10.1038/s41574-023-00851-2; Till Roenneberg, Karla V. Allebrandt, Martha Merrow, and Céline Vetter, "Social Jetlag and Obesity," *Current Biology* 22, no. 10 (May 2012): 939–43, doi.org/10.1016/j.cub.2012.03.038; and Sara Gamboa Madeira et al., "Social Jetlag, a Novel Predictor for High Cardiovascular Risk in Blue-Collar Workers Following Permanent Atypical Work Schedules," *Journal of Sleep Research* 30, no. 6 (December 2021): e13380, doi.org/10.1111/jsr.13380.

58 **Slightly more common than FASP:** This condition goes by many names, including delayed sleep phase disorder. Gian Carlo G. Parico et al., "The Human CRY1 Tail Controls Circadian Timing by Regulating Its Association with CLOCK:BMAL1," *Proceedings of the National Academy of Sciences of the United States of America* 117, no. 45 (November 2020): 27971–79, doi.org/10.1073/pnas.1920653117.

62 **Tommasi and his colleagues:** Alessio Gaggero and Denni Tommasi, "Time of Day and High-Stake Cognitive Assessments," *Economic Journal* 133, no. 652 (May 2023): 1407–29, doi.org/10.1093/ej/ueac090.

62 **Circadian variation across the day:** Carolyn B. Hines, "Time-of-Day Effects on Human Performance," *Journal of Catholic Education* 7, no. 3 (2004): 390–413, doi:10.15365/joce.0703072013.

63 **likelihood that we act morally:** Brian C. Gunia, Christopher M. Barnes, and Sunita Sah, "The Morality of Larks and Owls: Unethical Behavior Depends on Chronotype as Well as Time of Day," *Psychological Science* 25, no. 12 (December 2014): 2272–74, doi.org/10.1177/0956797614541989.

63 **"flash of illuminance":** Mareike B. Wieth and Rose T. Zacks, "Time of Day Effects on Problem Solving: When the Non-optimal Is Optimal," *Thinking & Reasoning* 17, no. 4 (2011): 387–401, doi.org/10.1080/13546783.2011.625663; and Janet Metcalfe

and David Wiebe, "Intuition in Insight and Noninsight Problem Solving," *Memory & Cognition* 15, no. 3 (May 1987): 238–46, doi.org/10.3758/BF03197722. Quoted in Daniel Pink, *When: The Scientific Secrets of Perfect Timing* (New York: Riverhead, 2018), 24.

63 **If you're more interested:** Shogo Sato et al., "Atlas of Exercise Metabolism Reveals Time-Dependent Signatures of Metabolic Homeostasis," *Cell Metabolism* 34, no. 2 (February 2022): 329–345.e8, doi.org/10.1016/j.cmet.2021.12.016.

64 **an athlete's performance:** Elise Facer-Childs and Roland Brandstaetter, "The Impact of Circadian Phenotype and Time since Awakening on Diurnal Performance in Athletes," *Current Biology* 25, no. 4 (February 2015): 518–22, doi.org/10.1016/j.cub.2014.12.036.

65 **McHill could finally tease apart:** Andrew W. McHill and Evan D. Chinoy, "Utilizing the National Basketball Association's COVID-19 Restart 'Bubble' to Uncover the Impact of Travel and Circadian Disruption on Athletic Performance," *Scientific Reports* 10, no. 1 (December 2020): 21827, doi.org/10.1038/s41598-020-78901-2.

66 **Mah collaborated with ESPN:** Baxter Holmes, "How Fatigue Shaped the NBA Season, and What It Means for the Playoffs," *ESPN*, April 10, 2018, espn.com/nba/story/_/id/23094298/how-fatigue-shaped-nba-season-means-playoffs.

67 **One study found MLB teams:** The researchers did not analyze game times and, therefore, couldn't tease out the impact of circadian timing. Alex Song, Thomas Severini, and Ravi Allada, "How Jet Lag Impairs Major League Baseball Performance," *Proceedings of the National Academy of Sciences of the United States of America* 114, no. 6 (January 2017): 1407–12, doi:10.1073/pnas.1608847114.

68 **When researchers considered chronotype:** Facer-Childs and Brandstaetter, "The Impact of Circadian Phenotype."

CHAPTER 4. RHYTHM AND BLUES

76 **Yet, despite the mice's blindness:** Clyde Keeler referred to "rodless" mice. Because cones would have been extremely difficult to discriminate with technology at the time, Ignacio Provencio noted that Keeler most likely meant photoreceptor-less mice—or mice deficient in both rods and cones. Clyde E. Keeler, "On the Occurrence in the House Mouse of Mendelizing Structural Defect of the Retina Producing Blindness," *Proceedings of the National Academy of Sciences of the United States of America* 12, no. 4 (1926): 255–58, doi:10.1073/pnas.12.4.255; and Clyde E. Keeler, "Iris Movements in Blind Mice," *American Journal of Physiology* 81, no. 1 (June 1927): 107–12, doi.org/10.1152/ajplegacy.1927.81.1.107.

76 **the idea of a third photoreceptor:** Some researchers prefer to call this the fifth photoreceptor, since there are three types of cone cells in the human eye.

76 **Foster and his team went:** M. S. Freedman et al., "Regulation of Mammalian Circadian Behavior by Non-rod, Non-cone, Ocular Photoreceptors," *Science* 284, no. 5413 (April 1999): 502–4, doi/10.1126/science.284.5413.502.

76 **Scientists, including Foster:** Farhan H. Zaidi et al., "Short-Wavelength Light Sensitivity of Circadian, Pupillary, and Visual Awareness in Humans Lacking an Outer Retina," *Current Biology* 17, no. 24 (December 2007): 2122–28, doi.org/10.1016/j.cub.2007.11.034.

77 **a "third eye" of sorts:** S. Doyle and M. Menaker, "Circadian Photoreception in Vertebrates," Cold Spring *Harbor Symposia on Quantitative Biology* 72, no. 1 (2007): 499–508, doi:10.1101/sqb.2007.72.003; Vincent M. Cassone, "Avian Circadian Organization: A Chorus of Clocks," *Frontiers in Neuroendocrinology* 35, no. 1 (January 2014): 76–88, doi:10.1016/j.yfrne.2013.10.002.

78 **Provencio named it melanopsin:** Ignacio Provencio et al., "Melanopsin: An Opsin in Melanophores, Brain, and Eye," *Proceedings of the National Academy of Sciences of the United States of America* 95, no. 1 (January 1998): 340–45, doi.org/10.1073/pnas.95.1.340.

78 **Other investigators followed:** D. M. Berson, F. A. Dunn, and M. Takao, "Phototransduction by Retinal Ganglion Cells That Set the Circadian Clock," *Science* 295, no. 5557 (February 2002): 1070–73, doi.org/10.1126/science.1067262; and S. Hattar, H. W. Liao, M. Takao, D. M. Berson, and K. W. Yau, "Melanopsin-Containing Retinal Ganglion Cells: Architecture, Projections, and Intrinsic Photosensitivity," *Science* 295, no. 5557 (February 2002): 1065–70, doi.org/10.1126/science.1069609.

82 **short wavelengths, around 480 nanometers:** Peak sensitivity of melanopsin is generally considered to be approximately 480 nanometers. However, many scientists use other estimates. A slightly longer peak of 490 nanometers, for example, may better account for the natural filtering by the retina in an average adult.

84 **Studies in children:** Kevin W. Houser, Lisa Heschong, and Richard Lang, "Buildings, Lighting, and the Myopia Epidemic," *LEUKOS: The Journal of the Illuminating Engineering Society* 19, no. 1 (2023): 1–3, doi.org/10.1080/15502724.2022.2141503.

84 **defend the skin:** As our largest organ and barrier to the outside world, the skin tends to better brace itself against all invaders, including bacteria, viruses, and pollutants during the daytime. At night, the skin becomes more permeable as it repairs and sheds cells. This is also when itching usually intensifies. Mary S. Matsui, Edward Pelle, Kelly Dong, and Nadine Pernodet, "Biological Rhythms in the Skin," *International Journal of Molecular Sciences* 17, no. 6 (June 2016): 801, doi:10.3390/ijms17060801.

84 **exposing a mouse to ultraviolet light:** Shobhan Gaddameedhi et al., "Control of Skin Cancer by the Circadian Rhythm," *Proceedings of the National Academy of Sciences of the United States of America* 108, no. 46 (September 2011): 18790–95, doi:10.1073/pnas.1115249108; Shobhan Gaddameedhi, Christopher P. Selby, Michael G. Kemp, Rui Ye, and Aziz Sancar, "The Circadian Clock Controls Sunburn Apoptosis and Erythema in Mouse Skin," *The Journal of Investigative Dermatology* 135, no. 4 (April 2015): 1119–27, doi:10.1038/jid.2014.508.

86 **She was a chronobiologist:** P. J. de Coursey, "Daily Light Sensitivity Rhythm in a Rodent," *Science* 131, no. 3392 (January 1960): 33–35, doi.org/10.1126/science.131.3392.33.

87 **The best predictor:** Gideon P. Dunster et al., "Daytime Light Exposure Is a Strong Predictor of Seasonal Variation in Sleep and Circadian Timing of University Students," *Journal of Pineal Research* 74, no. 2 (March 2023): e12843, doi.org/10.1111/jpi.12843.

91 **The first scientific case report:** Alfred J. Lewy et al., "Bright Artificial Light Treatment of a Manic-Depressive Patient with a Seasonal Mood Cycle," *American Journal of Psychiatry* 139, no. 11 (November 1982): 1496–98, doi.org/10.1176/ajp.139.11.1496.

92 **And Cook was the first:** Julian Sancton, *Madhouse at the End of the Earth: The* Belgica*'s Journey into the Dark Antarctic Night* (New York: Crown, 2021).

97 **Research has shown that exercise:** Dietmar Weinert and Denis Gubin, "The Impact of Physical Activity on the Circadian System: Benefits for Health, Performance and Wellbeing," *Applied Sciences* 12, no. 18 (September 2022): 9220, doi.org/10.3390/app12189220; and Ryan A. Martin and Karyn A. Esser, "Time for Exercise? Exercise and Its Influence on the Skeletal Muscle Clock," *Journal of Biological Rhythms* 37, no. 6 (December 2022): 579–92, doi.org/10.1177/07487304221122662.

97 **Because the Earth's rotation results:** Derk-Jan Dijk and Anne C. Skeldon, "Human

Sleep before the Industrial Era," *Nature* 527, no. 7577 (November 2015): 176–77, doi.org/10.1038/527176a.

97 **This internal temperature flux:** Ethan D. Buhr, Seung-Hee Yoo, and Joseph S. Takahashi, "Temperature as a Universal Resetting Cue for Mammalian Circadian Oscillators," *Science* 330, no. 6002 (October 2010): 379–85, doi.org/10.1126/science.1195262.

98 **The moon's gravitational pull:** C. Helfrich-Förster et al., "Women Temporarily Synchronize Their Menstrual Cycles with the Luminance and Gravimetric Cycles of the Moon," *Science Advances* 7, no. 5 (January 2021): eabe1358, doi.org/10.1126/sciadv.abe1358.

98 **Meanwhile, Horacio de la Iglesia:** Leandro Casiraghi et al., "Moonstruck Sleep: Synchronization of Human Sleep with the Moon Cycle under Field Conditions," *Science Advances* 7, no. 5 (January 2021): eabe0465, doi.org/10.1126/sciadv.abe0465.

98 **The idea that magnetic fields:** R. Wever, "The Effects of Electric Fields on Circadian Rhythmicity in Men," *Life Sciences and Space Research* 8 (1970): 177–87.

99 **Evidence suggests other environmental signals:** Musoki Mwimba et al., "Daily Humidity Oscillation Regulates the Circadian Clock to Influence Plant Physiology," *Nature Communications* 9 (October 2018): 4290, doi.org/10.1038/s41467-018-06692-2.

CHAPTER 5. DARK DAYS

103 **Glass windows eventually:** Shirin Hirsch and Andrew Smith, "A View through a Window: Social Relations, Material Objects and Locality," *Sociological Review* 66, no. 1 (January 2018): 224–40, doi.org/10.1177/0038026117724068.

103 **Between 1696 and 1851:** Meredith R. Conway, "And You May Ask Yourself, What Is That Beautiful House: How Tax Laws Distort Behavior through the Lens of Architecture," *Columbia Journal of Tax Law* 10, no. 2 (Summer 2019): 165–97, doi.org/10.7916/cjtl.v10i2.3468.

105 **"distorted" property owners' decisions:** Wallace E. Oates and Robert M. Schwab, "The Window Tax: A Case Study in Excess Burden," *Journal of Economic Perspectives* 29, no. 1 (Winter 2015): 163–80, doi.org/10.1257/jep.29.1.163.

108 **Warnings from the medical community:** *The Lancet*, February 22, 1845, 214–15.

109 **linked low light during the day:** The possibility remains that mental health problems drive unhealthy light and dark exposures. Experts even suggest it might go both ways, causing a vicious cycle as people stay indoors and cut off from natural cycles. Angus C. Burns et al., "Day and Night Light Exposure Are Associated with Psychiatric Disorders: An Objective Light Study in >85,000 People," *Nature Mental Health* 1, no. 11 (November 2023): 853–62, doi.org/10.1038/s44220-023-00135-8.

109 **animals living with daily shifts:** Rohan Nagare, Bernard Possidente, Sarita Lagalwar, and Mariana G. Figueiro, "Robust Light-Dark Patterns and Reduced Amyloid Load in an Alzheimer's Disease Transgenic Mouse Model," *Scientific Reports* 10, no. 1 (July 2020): 11436, doi.org/10.1038/s41598-020-68199-5.

111 **In 2022, a team:** Timothy M. Brown et al., "Recommendations for Daytime, Evening, and Nighttime Indoor Light Exposure to Best Support Physiology, Sleep, and Wakefulness in Healthy Adults," *PLoS Biology* 20, no. 3 (March 2022): e3001571, doi.org/10.1371/journal.pbio.3001571.

123 **Modern science supports her intuition:** R. S. Ulrich, "View through a Window May Influence Recovery from Surgery," *Science* 224, no. 4647 (April 1984): 420–21, doi.org/10.1126/science.6143402.

CHAPTER 6. BRIGHT NIGHTS

125 **Reflections off asphalt:** Researchers found a combination of cloud cover and snow cover increased the nighttime artificial skyglow in a suburb by a factor of 188. Andreas Jechow and Franz Hölker, "Snowglow—the Amplification of Skyglow by Snow and Clouds Can Exceed Full Moon Illuminance in Suburban Areas," in "Light Pollution Assessment with Imaging Devices," ed. Andreas Jechow, special issue, *Journal of Imaging* 5, no. 8 (August 2019): 69, doi.org/10.3390/jimaging5080069.

126 **That compounds to light pollution:** Christopher C. M. Kyba, Yiğit Öner Altıntaş, Constance E. Walker, and Mark Newhouse, "Citizen Scientists Report Global Rapid Reductions in the Visibility of Stars from 2011 to 2022," *Science* 379, no. 6629 (January 2023): 265–68, doi.org/10.1126/science.abq7781.

126 **That surge in short-wavelength light:** Alejandro Sánchez de Miguel, Jonathan Bennie, Emma Rosenfeld, Simon Dzurjak, and Kevin J. Gaston, "Environmental Risks from Artificial Nighttime Lighting Widespread and Increasing across Europe," *Science Advances* 8, no. 37 (September 2022): eabl6891, doi.org/10.1126/sciadv.abl6891.

128 **nearly one of every two homes:** Sean W. Cain et al., "Evening Home Lighting Adversely Impacts the Circadian System and Sleep," *Scientific Reports* 10, no. 1 (November 2020): 19110, doi.org/10.1038/s41598-020-75622-4.

128 **Artificial light at night has:** A small sample of this literature: Yong-Moon Mark Park et al., "Association of Exposure to Artificial Light at Night While Sleeping with Risk of Obesity in Women," *JAMA Internal Medicine* 179, no. 8 (August 2019): 1061–71, doi.org/10.1001/jamainternmed.2019.0571; and A. Green et al., "0029 Light Emitted from Media Devices at Night Is Associated with Decline in Sperm Quality," *Sleep* 43, no. Supplement_1 (April 2020): A12, doi.org/10.1093/sleep/zsaa056.028.

129 **rates and severity of COVID-19 infections:** The findings held after researchers accounted for other relevant factors, such as population density and wealth. Amedeo Argentiero, Roy Cerqueti, and Mario Maggi, "Outdoor Light Pollution and COVID-19: The Italian Case," *Environmental Impact Assessment Review* 90 (September 2021): 106602, doi.org/10.1016/j.eiar.2021.106602; and Yidan Meng, Vincent Zhu, and Yong Zhu, "Co-distribution of Light at Night (LAN) and COVID-19 Incidence in the United States," *BMC Public Health* 21 (August 2021), 1509doi.org/10.1186/s12889-021-11500-6.

129 **In a separate study:** Minjee Kim et al., "Light at Night in Older Age Is Associated with Obesity, Diabetes, and Hypertension," *Sleep* 46, no. 3 (March 2023), doi.org/10.1093/sleep/zsac130.

129 **Participants who felt like they:** Ivy C.Mason et al., "Light Exposure during Sleep Impairs Cardiometabolic Function," *Proceedings of the National Academy of Sciences of the United States of America* 119, no. 12 (March 2022): e2113290119, doi.org/10.1073/pnas.2113290119.

129 **A study using astronauts' photos:** Ariadna Garcia-Saenz et al., "Evaluating the Association between Artificial Light-at-Night Exposure and Breast and Prostate Cancer Risk in Spain (MCC-Spain Study)," *Environmental Health Perspectives* 126, no. 4 (April 2018): 047011, doi.org/10.1289/EHP1837.

129 **used satellite images and found:** Dong Zhang et al., "Associations between Artificial Light at Night and Risk for Thyroid Cancer: A Large US Cohort Study," *Cancer* 127, no. 9 (May 2021): 1448–58, doi.org/10.1002/cncr.33392.

133 **Wildlife also suffers:** Annika K. Jägerbrand and Kamiel Spoelstra, "Effects of Anthropogenic Light on Species and Ecosystems," *Science* 380, no. 6650 (June 2023): 1125–30, doi.org/10.1126/science.adg3173.

139 **In 1990, a federal court:** *Keenan v. Hall*, 83 F.3d 1083 (9th Cir. 1996).

139 **And, in 2018:** Charles A. Czeisler, "Housing Immigrant Children—the Inhumanity of Constant Illumination," *New England Journal of Medicine* 379, no. 2 (July 2018): e3, doi.org/10.1056/NEJMp1808450.

140 **People of color and low-income residents:** Shawna M. Nadybal, Timothy W. Collins, and Sara E. Grineski, "Light Pollution Inequities in the Continental United States: A Distributive Environmental Justice Analysis," *Environmental Research* 189 (October 2020): 109959, doi.org/10.1016/j.envres.2020.109959.

143 **One study found reductions:** David Mitre-Becerril, Sarah Tahamont, Jason Lerner, and Aaron Chalfin, "Can Deterrence Persist? Long-Term Evidence from a Randomized Experiment in Street Lighting," *Criminology & Public Policy* 21, no. 4 (November 2022): 865–91, doi.org/10.1111/1745-9133.12599.

146 **adopted strict national laws:** Martin Morgan-Taylor, "Regulating Light Pollution: More Than Just the Night Sky," *Science* 380, no. 6650 (June 2023): 1118–20, doi.org/10.1126/science.adh7723.

149 **A 359-page report:** War Department, *Control of Coastal Lighting in Anti-submarine Warfare*, no. 756 (Fort Belvoir, VA: Engineer Board, April 30, 1943): apps.dtic.mil/sti/pdfs/ADA954894.pdf.

CHAPTER 7. CLOCK SCRAMBLERS

154 **To Hurley's surprise:** Kayla D. Coldsnow, Rick A. Relyea, and Jennifer M. Hurley, "Evolution to Environmental Contamination Ablates the Circadian Clock of an Aquatic Sentinel Species," *Ecology and Evolution* 7, no. 23 (December 2017): 10339–49, doi.org/10.1002/ece3.3490.

155 **Many traditional contaminants:** A few selections from this research: Xiangming Hu et al., "Long-Term Exposure to Ambient Air Pollution, Circadian Syndrome and Cardiovascular Disease: A Nationwide Study in China," *Science of the Total Environment* 868 (April 2023): 161696, doi.org/10.1016/j.scitotenv.2023.161696; and Renate Kopp, Irene Ozáez Martínez, Jessica Legradi, and Juliette Legler, "Exposure to Endocrine Disrupting Chemicals Perturbs Lipid Metabolism and Circadian Rhythms," *Journal of Environmental Sciences* 62 (December 2017): 133–37, doi.org/10.1016/j.jes.2017.10.013.

155 **The timing of these exposures:** Jacqueline M. Leung and Micaela E. Martinez, "Circadian Rhythms in Environmental Health Sciences," *Current Environmental Health Reports* 7, no. 3 (September 2020): 272–81, doi.org/10.1007/s40572-020-00285-2.

157 **Caffeine affects both the circadian:** Tina M. Burke et al., "Effects of Caffeine on the Human Circadian Clock In Vivo and In Vitro," *Science Translational Medicine* 7, no. 305 (September 2015): 305ra146, doi.org/10.1126/scitranslmed.aac5125.

159 **Half of adults:** Shubhroz Gill and Satchidananda Panda, "A Smartphone App Reveals Erratic Diurnal Eating Patterns in Humans That Can Be Modulated for Health Benefits," *Cell Metabolism* 22, no. 5 (November 2015): 789–98, doi.org/10.1016/j.cmet.2015.09.005.

162 **Our fat-burning system:** Satchin Panda, *The Circadian Code: Lose Weight, Supercharge Your Energy, and Transform Your Health from Morning to Midnight* (New York: Rodale, 2018), 192.

164 **Mice who ate those unhealthy:** Megumi Hatori et al., "Time-Restricted Feeding without Reducing Caloric Intake Prevents Metabolic Diseases in Mice Fed a High-Fat Diet," *Cell Metabolism* 15, no. 6 (June 2012): 848–60, doi.org/10.1016/j.cmet.2012.04.019.

164 **In 2019, researchers from Salk:** Michael J. Wilkinson et al., "Ten-Hour Time-Restricted Eating Reduces Weight, Blood Pressure, and Atherogenic Lipids in Patients with Metabolic Syndrome," *Cell Metabolism* 31, no. 1 (January 2020): 92–104.e5, doi.org/10.1016/j.cmet.2019.11.004.

164 **In studies of mice:** Daniel S. Whittaker et al., "Circadian Modulation by Time-Restricted Feeding Rescues Brain Pathology and Improves Memory in Mouse Models of Alzheimer's Disease," *Cell Metabolism* 35, no. 10 (October 2023): 1704–1721.e6, doi.org/10.1016/j.cmet.2023.07.014.

165 **After designing and constructing feeders:** Victoria Acosta-Rodríguez et al., "Circadian Alignment of Early Onset Caloric Restriction Promotes Longevity in Male C57BL/6J Mice," *Science* 376, no. 6598 (May 2022): 1192–1202, doi.org/10.1126/science.abk0297.

166 **The findings so far have overthrown the paradigm:** Maria M. Mihaylova et al., "When a Calorie Is Not Just a Calorie: Diet Quality and Timing as Mediators of Metabolism and Healthy Aging," *Cell Metabolism* 35, no. 7 (July 2023): 1114–31, doi.org/10.1016/j.cmet.2023.06.008.

167 **the women who ate like queens:** Daniela Jakubowicz, Maayan Barnea, Julio Wainstein, and Oren Froy, "High Caloric Intake at Breakfast vs. Dinner Differentially Influences Weight Loss of Overweight and Obese Women," *Obesity* 21, no. 12 (December 2013): 2504–12, doi.org/10.1002/oby.20460.

171 **New mothers commonly:** Jennifer Hahn-Holbrook et al., "Human Milk as 'Chrononutrition': Implications for Child Health and Development," *Pediatric Research* 85, no. 7 (June 2019): 936–42, doi.org/10.1038/s41390-019-0368-x.

CHAPTER 8. MISMATCHED HOURS

177 **The unfortunate aptness of Boom:** If you're curious, my car's name was inspired by a group of Seahawks players nicknamed the "Legion of Boom."

188 **One team of economists estimated:** The researchers used statistical methods to account for potential confounders, such as the presence of high-earning metropolitan areas along the eastern seaboard. Osea Giuntella and Fabrizio Mazzonna, "Sunset Time and the Economic Effects of Social Jetlag: Evidence from US Time Zone Borders," *Journal of Health Economics* 65 (May 2019): 210–26, doi.org/10.1016/j.jhealeco.2019.03.007.

188 **The sudden change can botch:** Thomas Kantermann, Myriam Juda, Martha Merrow, and Till Roenneberg, "The Human Circadian Clock's Seasonal Adjustment Is Disrupted by Daylight Saving Time," *Current Biology* 17, no. 22 (November 2007): 1996–2000, doi.org/10.1016/j.cub.2007.10.025.

190 **More than 40 percent:** "Why Change?," Start School Later, accessed November 24, 2023, startschoollater.net/why-change.html.

191 **Teenagers' predilection to stay:** Till Roenneberg, *Internal Time: Chronotypes, Social Jet Lag, and Why You're So Tired* (Cologne: DuMont Buchverlag, 2012), 102.

192 **Starting as early as elementary:** Giulia Zerbini et al., "Lower School Performance in Late Chronotypes: Underlying Factors and Mechanisms," *Scientific Reports* 7 (June 2017): 4385, doi.org/10.1038/s41598-017-04076-y.

194 **One study found hospitalized shift:** Robert Maidstone et al., "Shift Work Is Associated with Positive COVID-19 Status in Hospitalised Patients," *Thorax* 76, no. 6 (June 2021): 601–6, thorax.bmj.com/content/76/6/601.

195 **Still, the conflicting results:** Aziz Sancar and Russell N. Van Gelder, "Clocks, Cancer, and Chronochemotherapy," *Science* 371, no. 6524 (January 2021): eabb0738, doi.org/10.1126/science.abb0738.

195 **Pregnant women working:** Danielle A. Clarkson-Townsend et al., "Maternal Circadian Disruption Is Associated with Variation in Placental DNA Methylation," *PloS One* 14, no. 4 (2019): e0215745, doi.org/10.1371/journal.pone.0215745.

195 **Dad, too, could transfer troubles:** Maximilian Lassi et al., "Disruption of Paternal Circadian Rhythm Affects Metabolic Health in Male Offspring via Nongerm Cell Factors," *Science Advances* 7, no. 22 (May 2021): eabg6424, doi.org/10.1126/sciadv.abg6424.

CHAPTER 9. GOODBYE, ALARM CLOCK

203 **In a photograph taken:** Much of Mary Smith's story, and that of the knocker-upper profession, is recounted in this paper: Arunima Datta, "Knocker Ups: A Social History of Waking Up in Victorian Britain's Industrial Towns," *Journal of Victorian Culture* 25, no. 3 (July 2020): 331–48, doi.org/10.1093/jvcult/vcaa013.

211 **Younger workers tend:** This section leans on an interview with Helen Burgess and her paper: Helen J. Burgess, Katherine M. Sharkey, and Charmane I. Eastman, "Bright Light, Dark and Melatonin Can Promote Circadian Adaptation in Night Shift Workers," *Sleep Medicine Reviews* 6, no. 5 (October 2002): 407–20, doi.org/10.1016/S1087-0792(01)90215-1.

212 **partially re-entrained workers:** Mark R. Smith, Louis F. Fogg, and Charmane I. Eastman, "A Compromise Circadian Phase Position for Permanent Night Work Improves Mood, Fatigue, and Performance," *Sleep* 32, no. 11 (November 2009): 1481–89, doi.org/10.1093/sleep/32.11.1481.

216 **Before-and-after data collected:** Gideon P. Dunster et al., "Sleepmore in Seattle: Later School Start Times Are Associated with More Sleep and Better Performance in High School Students," *Science Advances* 4, no. 12 (December 2018): eaau6200, doi.org/10.1126/sciadv.aau6200.

224 **Again, scientists fervently warn:** Till Roenneberg et al., "Why Should We Abolish Daylight Saving Time?," *Journal of Biological Rhythms* 34, no. 3 (June 2019): 227–30, doi.org/10.1177/0748730419854197.

224 **David Prerau, author,** David Prerau, "Advantages Abound with Changing Clocks Twice a Year," *The Sun* (Lowell, MA), March 12, 2023, lowellsun.com/2023/03/12/david-prerau-advantages-abound-with-changing-clocks-twice-a-year.

226 **Till Roenneberg and colleagues have:** Till Roenneberg, Eva C. Winnebeck, and Elizabeth B. Klerman, "Daylight Saving Time and Artificial Time Zones—a Battle between Biological and Social Times," *Frontiers in Physiology* 10 (August 2019): 944, doi.org/10.3389/fphys.2019.00944.

CHAPTER 10. LET THERE BE LIGHT, AND DARK

238 **studies led by Mariana Figueiro:** Mariana G. Figueiro et al., "The Impact of Daytime Light Exposures on Sleep and Mood in Office Workers," *Sleep Health* 3, no. 3 (June 2017): 204–15, doi.org/10.1016/j.sleh.2017.03.005.

239 **In a 2023 paper:** Martin Moore-Ede et al., "Lights Should Support Circadian Rhythms: Evidence-Based Scientific Consensus," *Frontiers in Photonics* 4 (2023): 1272934, doi.org/10.3389/fphot.2023.1272934.

240 **She described a cautionary case:** Food and Drug Administration, "Lamp's Labeling Found to be Fraudulent," *FDA Talk Paper* T86-69, September 10, 1986, cdn.centerforinquiry.org/wp-content/uploads/sites/33/2021/03/22170721/vita-lite_fraud_notice_1986.pdf.

247 **These small changes combined:** Robert Soler and Erica Voss, "Biologically Rele-

vant Lighting: An Industry Perspective," *Frontiers in Neuroscience* 15 (2021): 637221, doi.org/10.3389/fnins.2021.637221.

247 **Now slightly sleep deprived:** Leilah K. Grant et al., "Supplementation of Ambient Lighting with a Task Lamp Improves Daytime Alertness and Cognitive Performance in Sleep-Restricted Individuals," *Sleep* 46, no. 8 (August 2023): zsad096, doi.org/10.1093/sleep/zsad096.

247 **Models can now evaluate:** John Mardaljevic. "The Implementation of Natural Lighting for Human Health from a Planning Perspective." *Lighting Research & Technology* (London, England: 2001) 53, no. 5 (2021): 489–513. doi:10.1177/14771535211022145.

250 **Blue light and violet light could be:** D. Van Gilst et al., "Effects of the Neonatal Intensive Care Environment on Circadian Health and Development of Preterm Infants," *Frontiers in Physiology* 14 (August 2023): 1243162, doi.org/10.3389/fphys.2023.1243162.

253 **One study linked cataract removal:** Cecilia S. Lee et al., "Association between Cataract Extraction and Development of Dementia," *JAMA Internal Medicine* 182, no. 2 (February 2022): 134–41, doi.org/10.1001/jamainternmed.2021.6990.

255 **nursing homes in Wisconsin:** Leilah K. Grant et al., "Impact of Upgraded Lighting on Falls in Care Home Residents," *Journal of the American Medical Directors Association* 23, no. 10 (October 2022): 1698–1704.e2, doi.org/10.1016/j.jamda.2022.06.013.

CHAPTER 11. HACKING RHYTHMS

264 **She referred me to:** "Woman Who Drinks 6 Cups of Coffee Per Day Trying to Cut Down on Blue Light at Bedtime," *The Onion*, April 4, 2017, theonion.com/woman-who-drinks-6-cups-of-coffee-per-day-trying-to-cut-1819579770.

267 **Mattress companies, including:** Shahab Haghayegh et al., "Novel Temperature-Controlled Sleep System to Improve Sleep: A Proof-of-Concept Study," *Journal of Sleep Research* 31, no. 6 (December 2022): e13662, doi.org/10.1111/jsr.13662.

270 **The US Department of Defense reported:** Sarah Chabal, Rachel R. Markwald, Evan D. Chinoy, Joseph DeCicco, and Emily Moslener, *Personal Light Treatment Devices as a Viable Countermeasure for Submariner Fatigue* (Groton, CT: Naval Submarine Medical Research Laboratory, 2022), apps.dtic.mil/sti/pdfs/AD1166064.pdf.

273 **the 2020 Summer Olympics:** Due to the COVID-19 pandemic, the games were held in 2021.

275 **selecting more night owls:** Elise Facer-Childs and Roland Brandstaetter, "Circadian Phenotype Composition Is a Major Predictor of Diurnal Physical Performance in Teams," *Frontiers in Neurology* 6 (October 2015): 208, doi.org/10.3389/fneur.2015.00208.

275 **The average peak performance:** R. Lok, G. Zerbini, M. C. M. Gordijn, D. G. M. Beersma, and R. A. Hut, "Gold, Silver or Bronze: Circadian Variation Strongly Affects Performance in Olympic Athletes," *Scientific Reports* 10 (October 2020): 16088, doi.org/10.1038/s41598-020-72573-8.

278 **my visit to Palo Alto:** Renske Lok, Marisol Duran, and Jamie M. Zeitzer, "Moving Time Zones in a Flash with Light Therapy during Sleep," *Scientific Reports* 13 (September 2023): 14458, doi.org/10.1038/s41598-023-41742-w.

280 **The aim is to create:** Christopher Bettinger, "ADvanced Acclimation and Protection Tool for Environmental Readiness (ADAPTER)," Defense Advanced Research Projects Agency, accessed November 24, 2023, https://www.darpa.mil/program/advanced-acclimation-and-protection-tool-for-environmental-readiness.

CHAPTER 12. CIRCADIAN MEDICINE

282 **Morning is apparently also preferred:** Daniel H. Pink, *When: The Scientific Secrets of Perfect Timing* (New York: Riverhead Books, 2018), 54–55.

282 **accounting for daily rhythms:** A sampling of the literature: Michelle L. Gumz et al., "Toward Precision Medicine: Circadian Rhythm of Blood Pressure and Chronotherapy for Hypertension—2021 NHLBI Workshop Report," *Hypertension* 80, no. 3 (March 2023): 503–22, doi.org/10.1161/hypertensionaha.122.19372; Yihao Liu et al., "The Impact of Circadian Rhythms on the Immune Response to Influenza Vaccination in Middle-Aged and Older Adults (IMPROVE): A Randomised Controlled Trial," *Immunity & Ageing* 19, no. 1 (October 2022): 46, doi.org/10.1186/s12979-022-00304-w; and Maurizio Cutolo, "Glucocorticoids and Chronotherapy in Rheumatoid Arthritis," *RMD Open* 2, no. 1 (January 2016): e000203, doi.org/10.1136/rmdopen-2015-000203.

283 **Levels of circulating tumor:** Zoi Diamantopoulou et al., "The Metastatic Spread of Breast Cancer Accelerates during Sleep," *Nature* 607, no. 7917 (July 2022): 156–62, doi.org/10.1038/s41586-022-04875-y.

284 **In 1869, Hyde Salter:** Hyde Salter, "On the Treatment of Asthma by Belladonna," *The Lancet* 93, no. 2370 (January 30, 1869): 152–53, doi.org/10.1016/S0140-6736(02)65754-X.

284 **More than a century:** Frank A. J. L. Scheer et al., "The Endogenous Circadian System Worsens Asthma at Night Independent of Sleep and Other Daily Behavioral or Environmental Cycles," *Proceedings of the National Academy of Sciences of the United States of America* 118, no. 37 (2021): e2018486118, doi.org/10.1073/pnas.2018486118.

285 **fluctuating with the seasons:** Ludovic S. Mure et al., "Diurnal Transcriptome Atlas of a Primate across Major Neural and Peripheral Tissues," *Science* 359, no. 6381 (February 2018): doi.org/10.1126/science.aao0318.

285 **a lot of these rhythmic:** Marc D. Ruben, David F. Smith, Garret A. FitzGerald, and John B. Hogenesch, "Dosing Time Matters," *Science* 365, no. 6453 (August 2019): 547–49, doi.org/10.1126/science.aax7621.

285 **Research suggests that the majority:** Timed treatments are most relevant for drugs formulated to be short-acting, or with a half-life of less than around eight hours.

285 **Pertinent to my surgery:** H. Al-Waeli et al., "Chronotherapy of Non-steroidal Anti-inflammatory Drugs May Enhance Postoperative Recovery," *Scientific Reports* 10 (January 2020): 468, doi.org/10.1038/s41598-019-57215-y.

286 **Researchers discovered that timing mattered:** A few examples: Guy Hazan et al., "Biological Rhythms in COVID-19 Vaccine Effectiveness in an Observational Cohort Study of 1.5 Million Patients," *Journal of Clinical Investigation* 133, no. 11 (June 2023): e167339, doi.org/10.1172/jci167339; Candace D. McNaughton et al., "Diurnal Variation in SARS-CoV-2 PCR Test Results: Test Accuracy May Vary by Time of Day," *Journal of Biological Rhythms* 36, no. 6 (December 2021): 595–601, doi.org/10.1177/07487304211051841; Michael J. McCarthy, "Circadian Rhythm Disruption in Myalgic Encephalomyelitis/Chronic Fatigue Syndrome: Implications for the Post-acute Sequelae of COVID-19," *Brain, Behavior, & Immunity—Health* 20 (March 2022): 100412, doi.org/10.1016/j.bbih.2022.100412.

286 **jabs with an mRNA vaccine:** Researchers found this midday window to be optimal in a large retrospective study of mRNA COVID vaccination. But not all vaccine studies reach the same conclusion. For example, protein antigen vaccines and vac-

cines against influenza may peak in effectiveness at different times of day. Different biological pathways and different targets likely require different timing.

288 **We now know the microscopic:** Filipa Rijo-Ferreira et al., "The Malaria Parasite Has an Intrinsic Clock," *Science* 368, no. 6492 (May 2020): 746–53, doi.org/10.1126/science.aba2658.

291 **Circadian rhythm disruption and neurodegeneration:** Erik S. Musiek, David D. Xiong, and David M. Holtzman, "Sleep, Circadian Rhythms, and the Pathogenesis of Alzheimer Disease," *Experimental & Molecular Medicine* 47, no. 3 (March 2015): e148, doi.org/10.1038/emm.2014.121.

292 **Research suggests that daytime:** Nathaniel P. Hoyle et al., "Circadian Actin Dynamics Drive Rhythmic Fibroblast Mobilization during Wound Healing," *Science Translational Medicine* 9, no. 415 (November 2017): doi.org/10.1126/scitranslmed.aal2774.

293 **His data collected from neurons:** Kayla E. Rohr and Michael J. McCarthy, "The Impact of Lithium on Circadian Rhythms and Implications for Bipolar Disorder Pharmacotherapy," *Neuroscience Letters* 786 (August 2022): 136772, doi.org/10.1016/j.neulet.2022.136772.

294 **This discovery has led:** Zhen Dong et al., "Targeting Glioblastoma Stem Cells through Disruption of the Circadian Clock," *Cancer Discovery* 9, no. 11 (November 2019): 1556–73, doi.org/10.1158/2159-8290.CD-19-0215.

295 **They tested their theory:** Anna R. Damato et al., "Temozolomide Chronotherapy in Patients with Glioblastoma: A Retrospective Single-Institute Study," *Neuro-oncology Advances* 3, no. 1 (January–December 2021): vdab041, doi.org/10.1093/noajnl/vdab041.

295 **Medical schools pay scant:** Julia M. Selfridge, Kurtis Moyer, Daniel G. S. Capelluto, and Carla V. Finkielstein, "Opening the Debate: How to Fulfill the Need for Physicians' Training in Circadian-Related Topics in a Full Medical School Curriculum," *Journal of Circadian Rhythms* 13 (November 2015): 7, doi.org/10.5334/jcr.ah.

296 **Lo decided to dig:** Elga Esposito et al., "Potential Circadian Effects on Translational Failure for Neuroprotection," *Nature* 582, no. 7812 (June 2020): 395–98, doi.org/10.1038/s41586-020-2348-z.

298 **In one analysis of drug:** Marc D. Ruben et al., "A Large-Scale Study Reveals 24-h Operational Rhythms in Hospital Treatment," *Proceedings of the National Academy of Sciences of the United States of America* 116, no. 42 (October 2019): 20953–58, doi.org/10.1073/pnas.1909557116.

299 **Carole Godain remembers:** Lynne Peeples, "Medicine's Secret Ingredient—It's in the Timing." *Nature* 556, no. 7701 (2018): 290–92. doi:10.1038/d41586-018-04600-8. Further correspondence with Godain in 2023.

299 **Lévi, also a chronobiologist:** Francis Lévi, Rachid Zidani, and Jean-Louis Misset, "Randomised Multicentre Trial of Chronotherapy with Oxaliplatin, Fluorouracil, and Folinic Acid in Metastatic Colorectal Cancer," *The Lancet* 350, no. 9079 (September 1997): 681–86, doi.org/10.1016/S0140-6736(97)03358-8.

CHAPTER 13. EXTENDED HOURS

304 **Of course, life on Mars:** Rujia Luo, Yutao Huang, Huan Ma, and Jinhu Guo, "How to Live on Mars with a Proper Circadian Clock?," *Frontiers in Astronomy and Space Sciences* 8 (January 2022): 796943, doi.org/10.3389/fspas.2021.796943.

307 **harness and hack the clocks:** The following section leans on lessons from several researchers, including Alex Webb (University of Cambridge), Joanna Chiu (University

of California, Davis), Mark Rea (Mount Sinai), David Gadoury (Cornell University), Jose Pruneda-Paz (University of California, San Diego), Kathleen Greenham (University of Minnesota), Carlos Hotta (University of São Paulo), and Antony Dodd (John Innes Centre). A couple of in-depth review papers on the subject: Gareth Steed, Dora Cano Ramirez, Matthew A. Hannah, and Alex A. R. Webb, "Chronoculture, Harnessing the Circadian Clock to Improve Crop Yield and Sustainability," *Science* 372, no. 6541 (April 2021): eabc9141, doi.org/10.1126/science.abc9141; and Carlos Takeshi Hotta, "From Crops to Shops: How Agriculture Can Use Circadian Clocks," *Journal of Experimental Botany* 72, no. 22 (December 2021): 7668–79, doi.org/10.1093/jxb/erab371.

308 **The political geography:** Santiago Mora-García and Marcelo J. Yanovsky, "A Large Deletion within the Clock Gene *LNK2* Contributed to the Spread of Tomato Cultivation from Central America to Europe," *Proceedings of the National Academy of Sciences of the United States of America* 115, no. 27 (June 2018): 6888–90, doi.org/10.1073/pnas.1808194115.

311 **But if you artificially place:** Danielle Goodspeed et al., "*Arabidopsis* Synchronizes Jasmonate-Mediated Defense with Insect Circadian Behavior," *Proceedings of the National Academy of Sciences of the United States of America* 109, no. 12 (February 2012): 4674–77, doi.org/10.1073/pnas.1116368109.

312 **farmers need to spray:** Fiona E. Belbin et al., "Plant Circadian Rhythms Regulate the Effectiveness of a Glyphosate-Based Herbicide," *Nature Communications* 10 (August 2019): 3704, doi.org/10.1038/s41467-019-11709-5.

312 **Reducing application of Roundup:** Who.int, "IARC Monographs Volume 112: Evaluation of Five Organophosphate Insecticides and Herbicides," accessed February 5, 2024, https://www.iarc.who.int/news-events/iarc-monographs-volume-112-evaluation-of-five-organophosphate-insecticides-and-herbicides.

315 **The clock can keep ticking:** John D. Liu et al., "Keeping the Rhythm: Light/Dark Cycles during Postharvest Storage Preserve the Tissue Integrity and Nutritional Content of Leafy Plants," *BMC Plant Biology* 15 (March 2015): 92, doi.org/10.1186/s12870-015-0474-9.

319 **maximally harvest food:** Matteo Camporese and Majdi Abou Najm, "Not All Light Spectra Were Created Equal: Can We Harvest Light for Optimum Food-Energy Cogeneration?," *Earth's Future* 10, no. 12 (December 2022): e2022EF002900, doi.org/10.1029/2022EF002900.

Index